D0874238

MECHANISM, LIFE AND
PERSONALITY

MECHANISM, LIFE AND PERSONALITY

AN EXAMINATION OF THE MECHANISTIC THEORY OF LIFE AND MIND

BY

J. S. HALDANE, M.D., LL.D., F.R.S.

FELLOW OF NEW COLLEGE, UNIVERSITY OF OXFORD

SECOND EDITION

GREENWOOD PRESS, PUBLISHERS
WESTPORT, CONNECTICUT

The Library of Congress has catalogued this publication as follows:

Library of Congress Cataloging in Publication Data

Haldane, John Scott, 1860–1936.
   Mechanism, life, and personality.

   Reprint of the 1923 ed.
   1. Life. 2. Biology. 3. Personality.
I. Title.
B1645.H33M44 1973   128'.5   72–7966
ISBN 0–8371–6557–1

B1645
H33
M44
  1973

~~128~~

~~1158~~

Originally published in 1923
by E. P. Dutton and Company, New York

First Greenwood Reprinting 1973

Library of Congress Catalogue Card Number 72-7966

ISBN 0-8371-6557-1

Printed in the United States of America

PREFACE TO FIRST EDITION

THIS book consists of four lectures which were delivered in the Physiological Laboratory of Guy's Hospital, during May of this year, as a London University course for senior students. They are reproduced in the form of their delivery, after careful revision, in which I have been much aided by the criticisms and suggestions of my friend Professor J. T. Wilson, F.R.S.

Philosophical readers who may have chanced to see an essay by my brother and myself on 'The Relations of Philosophy to Science' in *Essays in Philosophical Criticism,* published in 1883, will recognise in these lectures a development of the ideas put forward in that essay. In a presidential address which I delivered in 1908 before the Physiological Section of the British Association, and in other scattered papers, the same line of argument in relation to the aims of biology and its position among the sciences was followed out in certain directions. The lectures now published represent an attempt at a more comprehensive treatment of the subject.

The time is now more than ripe for bringing the great biological movement of the nineteenth century into definite relation with the main stream of human thought; and these lectures form a contribution towards the fulfilment of this task.

PREFACE TO SECOND EDITION

IN the present edition I have entirely recast the fourth lecture, which deals with Personality. The subject is of supreme human interest, but proportionally difficult. Further reading and thought have brought to me additional light upon it, and I trust that in its new form the fourth lecture will be correspondingly clearer to others.

Except for one or two slight emendations the other three lectures are unaltered. If any readers wish to follow out in detail the application in physiology of the ideas formulated in the third lecture, I may perhaps refer them to my Silliman Lectures (now in the Press) on the Physiology of Respiration.

OXFORD.

CONTENTS

LECTURE I

LECTURE II

LECTURE III

LECTURE IV

LECTURE I

THE aim of the first two of these lectures is to examine the hypothesis that living organisms may be regarded as conscious or unconscious physical and chemical mechanisms, and can be satisfactorily investigated from this standpoint. In this first lecture I shall endeavour to state, as well as I can, the case for what may be called, in the absence of a better expression, the mechanistic theory of life.

The researches of countless investigators have established with practical unanimity certain very fundamental facts with regard to living organisms. One of these is that the matter of which the bodies of organisms are found by analysis to be composed consists of the same chemical elements as are found outside the body, and that no new matter is formed in the body, or disappears from it. All the matter which is found in the body, or which passes from it, can be accounted for by what is taken up from the environment. Of the particular chemical substances, moreover, which have been found in the body a large and ever increasing number can be formed artificially outside it, and there is no reason for believing that any ultimate difficulty will be experienced in artificially forming any of the chemical substances which have been discovered, or are ever likely to be discovered, within the body.

Another fundamental fact is that the whole of the energy which is liberated in the body, whether as heat, mechanical work, or in other forms, can be traced to sources outside the body. The actual external sources of energy in the living body were first pointed out in general terms by Mayer more than sixty years ago, and the exact investigations of subsequent physiologists have completely verified his general conclusions.

The two great physical laws of conservation of matter and conservation of energy can thus be extended with apparently rigorous accuracy to all living organisms, including human beings. From this it may be inferred that, however complex may be the changes involved in organic activity, they are nothing but changes in a material system. As yet we are far from being able to trace this system and its changes completely; but the main outlines are clear, and the gradual filling in of details can only be a matter of time, though we shall probably never succeed in completely filling in all the details.

It is true that, in the case of at least the higher organisms, consciousness accompanies some of the material changes in the living body, whereas consciousness is not known to accompany material changes in the inorganic world. Of this mysterious accompaniment we can of course give no physical account. Possibly consciousness accompanies all material change; but whether or not this is the case, consciousness seems to make no difference in the end to the physical and chemical chain of events. It might, conceivably, be a form of energy, generated under certain unknown conditions in active protoplasm; but, if so, it is only one form of energy, generated from and immediately passing back into other forms. Cut off the oxygen supply to the brain, and consciousness ceases within a few seconds. It makes no difference to the energy balance of the body whether an animal is conscious or not; and it seems simplest, in the present state of knowledge, to treat consciousness as an accompaniment, not altering in any way the physical and chemical changes which it accompanies, of certain intra-protoplasmic changes.

The mechanistic theory may be supported by another set of considerations, which appeal very strongly to many scientific investigators. In all biological investigations we are investigating either structure or activity; and when we come to details we find that the structure is physical and chemical structure, and the activity physical and chemical activity. Hence biology can be nothing but the physics and chemistry of organisms.

This argument may be illustrated in almost infinite detail. Let us suppose that we are engaged in anatomical investigation. With the scalpel, microtome, microscope, fixing and staining reagents and other physical or chemical apparatus we separate out or render distinguishable the details of structure. But these details are only details of form, size, colour, physical or chemical characters, and spatial relations to other parts. The facts we ascertain are physical and chemical facts; and the methods we use are physical and chemical methods.

If, on the other hand, we are making a physiological investigation, the same is no less true. If we are investigating secretion we are measuring the mass or volume of the substances secreted, or their chemical composition, or perhaps their osmotic pressure, or concentrations of ions in them. If we are investigating muscular contraction we are measuring the rate and extent of the contraction, or the accompanying heat production or chemical or electromotive phenomena. The phenomenon which we observe is always some physical or chemical change. The methods we use are physical and chemical methods, and the resulting facts are consequently physical and chemical facts.

From the very nature of its methods and

facts, therefore, biology can apparently be nothing but applied physics and chemistry. When we come to matters of detail it appears that biology does nothing, and can do nothing, but ask physical and chemical questions, and obtain physical and chemical answers, in so far as the answers are more than partial answers. In tracing, say, the development of an organism, or investigating the process of secretion, we, of course, obtain, in the present state of knowledge, only partial answers. We may not be able to give a complete statement in physical and chemical terms of the processes of development or of secretion; but it remains true, nevertheless, that the facts we are ascertaining are apparently nothing but physical and chemical facts, however imperfectly defined. Beyond these facts all is more or less empty speculation, which scientific men would probably do well to avoid.

For further support the mechanistic theory can appeal to the actual history of biology, and particularly of physiology. Apart from the constant controversies (to which reference will be made shortly) between mechanists and non-mechanists, the history of physiology displays uninterrupted progress in the successful application of physical and chemical methods to physiological problems. To take only a few examples, the principles of mechanics were applied by Borelli to elucidate the action of the muscles on the limbs. Kepler applied the principles of optics to the action of the eye in vision. Harvey established and analysed the facts relating to circulation of the blood by the application of purely physical observation and reasoning. As a result of the great advance in chemistry at the end of the eighteenth century the fundamental facts with regard to respiration and its relation to nutrition and animal heat were discovered; and since then the application of chemical methods to physiological problems has been continued with unbroken success. In more recent times the advance of physical chemistry has placed new and powerful weapons in the hands of physiologists. It is doubtless the case that the application of chemistry to physiology has also shown us that the chemistry of life is far more complex than was formerly suspected. This, however, cannot legitimately be used as an argument against the mechanistic theory : for the progress which has been made by applying chemical methods is solid progress, to which there has been no set-back, though the goal has turned out to be further away than appeared to be the case about the middle of last century, at the time of the general movement among physiologists towards the mechanistic theory. It is, perhaps, on the chemical side that physiology has made its most striking advances in recent times; but on the physical side the advance has also been continuous.

We may now turn to some of the arguments which have been used in favour of a non-mechanistic interpretation of biological facts. Perhaps the most striking fact with regard to physiological phenomena is the evidence they present of activity co-ordinated in such a manner as to conduce towards the survival of either the individual or the species. Co-ordination of a similar striking kind is not found anywhere outside the organic world; and the mere existence of this co-ordination has been taken as strong evidence for the presence in living organisms of some co-ordinating influence apart from blind physical and chemical forces.

The reply to this argument is that many of the mechanisms by which co-ordination is brought about have already been discovered, and that every year more is being discovered about them. Descartes, in his writings about the nervous system, was the first to point the way in this line of discovery. He suggested nervous mechanisms, by means of which afferent stimuli and muscular responses are co-ordinated; and since his time the theory that the nervous system is at bottom nothing more than a complex system of reflex mechanisms has been experimentally verified in many directions, and has now become a generally accepted physiological doctrine. The reflex actions associated with consciousness are evidently so complex that their gradual analysis may take generations of research work. We have also to bear in mind that many forms of nervous activity are excited or influenced by chemical stimuli acting, not on peripheral afferent nerves, but on the central nervous system. Thus the nervous mechanism by which the breathing is so regulated as to bring to the lungs just the requisite quantity of air, in spite of rapid and enormous changes in the volume of air required, is actuated by chemical stimuli conveyed to the respiratory centre by the blood which bathes it. The nervous or other mechanisms by means of which the activity of the heart and the distribution of blood in different parts of the body are properly regulated have also been to a large extent elucidated.

It is not, however, in connection with the nervous system only that co-ordinating mechanisms have been discovered. It has been known for long that chemical substances produced by the activity of one organ are conveyed by the blood to other organs, in which they excite activity of such a nature as to maintain the normal structure, composition and functional activity of the body as a whole. Thus the secretory action of the kidneys varies according to the nature and amount of the various substances discharged into the blood, or withdrawn from it, by other organs. If, for instance, an excess of water is taken up from the intestines, it is rapidly got rid of through increased excretion by the kidneys: while diminished excretion balances excessive withdrawal of water from the blood, as by sweating. The kidneys respond in a similar manner to excess of urea, sugar, inorganic salts, acids, or alkalies, etc., the result being that the normal composition of the blood is kept constant within small limits. Another similar case is that of the liver, which forms or discharges various important substances in response to chemical stimuli conveyed through the blood. Physiological investigation is constantly discovering new cases in which the activity of one organ is excited by the chemical products of activity in other organs. In all these cases the response follows the stimulus, the cause the effect, with unerring accuracy. Apart altogether from the nervous system, therefore, the body is full of mechanisms for the co-ordination and control of functional activity. The nervous system, indeed, has come in recent years to appear much less important than was formerly supposed in the co-ordination of physiological activity ; and it seems to be only in cases where very rapid control is required that the nervous system plays an important part.

It is not merely in the case of functional activity that chemical stimuli play an important part: for evidence is now steadily accumulating that the co-ordination of growth and maintenance in different parts of the body is dependent on the action of chemical substances conveyed from part to part by the blood and lymph circulation, or by simple diffusion from one contiguous part to another. Some of the most striking evidence in this direction has been afforded by the extraordinary effect on growth and nutrition which is produced by removal or disease of such organs as the thyroid gland or the pituitary body. Absence of the thyroid completely arrests growth, and influences in other ways the normal structure of the body; but these effects can be annulled by the administration of an extract of the thyroid from another animal. Hence even such a very small organ as the thyroid produces chemical substances which excite and control the nutrition of other parts. Another well-known case is the influence of changes in the generative organs on the growth and nutrition of distant parts of the body. Although the evidence is, owing to the great chemical and experimental difficulties, accumulating but slowly at present, we cannot doubt that the growth and maintenance of every part of a living organism is controlled by the chemical stimuli derived from other parts, or from the environment.

There thus appears to be no difficulty in regarding a living organism as a complex system of physico-chemical mechanisms, each of which is controlled by the rest in such a way that the normal structure and activity of the organism is, under ordinary conditions, maintained. We can, moreover, verify the existence of these mechanisms, one by one, by exact experiment; and their actual existence has already been verified in a large number of instances. Hence the co-ordination of growth and activity, as found in living organisms, presents, in itself, no real difficulties for the mechanistic theory.

We have still to account for the existence of physico-chemical structures possessing the marvellous characteristics of living organisms —so different from anything found in the inorganic world. Here the genius of Charles Darwin has provided an explanation in harmony with the mechanistic theory. No type of organism can survive in which the physiological mechanisms are not so constituted and so co-ordinated that the organism is capable of completing a life-cycle culminating in the transmission of a similar structure to its descendants. Those organisms and their descendants which have varied in structure in a direction which gives them a smaller probability of surviving will become extinct in the fierce struggle for existence which prevails everywhere among living organisms. On the other hand, those which have varied in such a direction as to give them a greater probability of surviving will increase and multiply, displacing the inferior types. Hence by the blind operation of natural selection through countless ages higher and higher types of organism will be produced.

The wonderful complexity, accuracy, and co-ordination of the physiological mechanisms found within the bodies of living organisms can thus be accounted for on purely mechanical principles. All that it is necessary to assume is that organisms transmit their structure to their descendants, but in such a manner as to permit of slight variations, on which natural selection then acts in such a way that the structure is, generation by generation, more and more perfected and elaborated.   With this perfecting of the structure which makes for the survival of the species there, of course, goes hand in hand the elaboration of the mechanism needed for securing the inheritance of the more perfect structure.

On the essential mechanism of hereditary transmission, recent investigation, and particularly the work of Weismann, has thrown a flood of light. The germ-plasm, from which parent and offspring are alike developed, is, we have now reason to believe, passed on qualitatively unaltered, or almost unaltered, from parent to offspring. It is constantly increasing in amount by a simple process of growth or quantitative increase; but as regards chemical structure and composition it remains the same. The offspring resembles the parent for the simple reason that both originate from material which is qualitatively identical, and which is placed originally in qualitatively the same physical and chemical environment. This being so, the only possible result is that the offspring resembles the parent.

The actual origin of living organisms from inorganic matter is still wrapt in mystery. We have not yet succeeded in producing in the laboratory anything really resembling any known living organism. This, however, is not to be wondered at, since all the organisms which we at present recognise as such are now known to be descended through count-less generations from their first progenitors. Throughout these countless generations natural selection has continuously acted, gradually evolving a complexity of structure which we can never hope to imitate by laboratory experiments.

It is possible, nevertheless, to imagine how life may have originated. We can now form in the laboratory carbon compounds with protein characters which approach in some degree in complexity to the natural protein compounds found in protoplasm ; and we may assume that the first step to-wards the origin of life was the formation by natural means of some variety of protein material. The complexity of the protein molecule is enormous, and its capacity of entering into chemical or physical combinations with other molecules is so great that in it we seem to see the possibilities of the evolution of something possessing the primitive characteristics of life. We can imagine such molecules combining by aggregation until the aggregate becomes unstable and divides into two aggregates, each possessing all the original properties of the first, and being capable of similarly growing and dividing. If the conditions are such that this process can go on continuously, and that new protein can constantly be produced, as is actually the case in vegetable tissues, natural selection will soon come into play, and the evolution of a less imperfect, and more definite and complex, form of organism will gradually take place.

A further part of the case for the mechanistic theory consists in criticisms of alternative theories. The traditional opponents of the mechanistic theory may be classed as either vitalists or animists. The former maintain that the behaviour of living organisms is so peculiar, and so different from anything met with in the inorganic world, that we must assume that the processes occurring in living organisms are guided and determined by some non-physical factor best known as the 'vital force' or 'vital principle,' but in recent times appearing under various other names. This factor is assumed to act blindly and unconsciously, and to be in action at all parts of the body. The animists also assume a non-physical factor; and this they identify with the soul, which is supposed, however, to act sub-consciously. For the purposes of the present argument the animists may, perhaps, be classified along with the vitalists, as the differences between them are for the present purpose very small.

Although outspoken vitalism has now become somewhat exceptional among biologists, or physiologists, and the prevailing opinion is in favour of the mechanistic theory, yet vitalism in a modified form seems to be constantly cropping up in biological literature. The reasons for this are, in part at least, fairly evident. We have only to look back to the writings of some of the leading advocates of mechanistic ideas in biology to see that in many instances simple mechanical theories of various physiological processes were put forward without sufficient experimental evidence of their correctness. Subsequent investigation has often shown that these theories were incorrect, and this result has seemed to be a justification for returning to the prevalent vitalistic ideas of earlier times.

As an instance of this we may take Ludwig's early theory as to the process by which urine is secreted. He supposed that urine is separated from the blood in the glomeruli of the kidney by a process of simple filtration, and that the dilute liquid thus formed is concentrated during its passage down the tubules by a process of osmosis, in which water passes from the tubules back into the blood, leaving the fully formed urine as a relatively concentrated liquid which passes out towards the ureter. This theory turned out, when experimentally tested, to be far from correct. It is the cells lining the tubules of the kidney which are the active agents in the secretion of urine; and they act against osmotic pressure, picking out from the blood, and forcing into the ducts of the tubules the various constituents of urine. Such a mode of statement seems to approximate very closely to vitalism, although, as already pointed out, we have only to presuppose the existence of suitable intra-cellular mechanisms in the cells lining the tubules in order to fit the observed facts with the mechanistic theory of life.

As another instance we may take the case of oxidation processes occurring in the living body. It was formerly believed that physiological oxidation is a comparatively simple process, varying in rate, like ordinary oxidations occurring outside the body, with the supply of oxygen on the one hand, and of oxidisable food-material on the other. When, however, the matter came to be investigated by Pflüger and others, it was found that this was not the case: for the supply of oxygen and of food-material could be varied within wide limits without altering the rate of oxidation. Hence it was concluded that the living cell regulates its own oxidation processes, and, indeed, all chemical processes occurring within it. This appears to be a very vitalistic form of conclusion: for some independent factor seems to be postulated which regulates the chemical processes occurring within cells. From the standpoint of the mechanistic theory, however, all that need be assumed is that the chemical processes occurring within living cells are far more complex than was at one time supposed. For one thing, intra-cellular ferments or enzymes seem to be largely concerned in intra-cellular changes. If, therefore, oxidations and other intra-cellular chemical processes are dependent on the liberation of enzymes, we can readily understand why these processes do not increase or diminish in the same manner as the simpler processes commonly met with outside the living body. By discovering the nature of the ferments concerned, and the conditions determining their liberation and mode of action, we may hope to be able to give a complete physico-chemical explanation of all the phenomena. In any case, this seems to be the only line of investigation likely to give fruitful results, and nothing could be more futile than to rest content with the theory that some mysterious agency is at work, the nature of which is beyond physical and chemical investigation.

It is only what might be expected on the Darwinian theory of natural selection that the physico-chemical mechanisms within the living body should turn out to be extremely complex, since these mechanisms are the product of natural selection acting through countless generations; and it is folly to go back to vitalism for the mere reason that living organisms do actually turn out to be very complex.

We must now examine the nature of the assumption made by the vitalists, and consider to what extent it corresponds with observed facts. Vitalism raises no objection to physical and chemical explanations applied in what is considered their proper place outside the intimate vital processes of living organisms. It assumes, however, that these intimate processes are guided or controlled by an influence which is manifested only in living organisms, and which acts in a manner wholly different from anything known in the inorganic world. The reason for making this assumption is that living organisms seem everywhere to present evidence of an autonomy of their own. They seem to go their own way, and pursue their own ends, in spite of all kinds of disturbing conditions. In a physical and chemical environment which is constantly changing in one respect or another, they develop or maintain certain characteristic peculiarities of structure and activity, and pursue a characteristic life-history. Mechanisms which would cause them to do this under the very varied conditions which they experience seem hardly conceivable. Hence we must assume in the case of each organism the presence of some influence which is constantly guiding in the right direction the otherwise blind physical and chemical processes within the living organism. This assumed influence is best known under the name of 'vital force'; but such a name is evidently unsuitable, as the word 'force' is usually employed to designate physical and chemical phenomena of a very different character. The expressions 'vital principle,' or 'entelechy' (proposed recently by Driesch) are more suitable.

The manner in which the mechanistic theory endeavours to meet the objections of the vitalists has already been indicated, but will be examined more closely in the second lecture. Meanwhile let us see what are the objections to vitalism.

Assuming the vitalistic theory, the simplest observations show us that the action of the vital principle is dependent on physical and chemical conditions of environment, and on the physical and chemical structure of the organism. If, for instance, the temperature is too high or too low, or if the supply of oxygen, or even of some inorganic salt, is cut off, all characteristic signs of life soon cease. The same is the case if the finer structure of the living material has been seriously damaged by physical or chemical agents. It is thus only under certain limiting physical and chemical conditions that the vital principle can manifest itself. Provided, however, that these limiting conditions are present there appears, at first sight, to be a wide field for the manifestation of the influence of the vital principle. If, for instance, a certain minimum of oxygen or of food-material is present, the organism selects just what it requires for its growth and maintenance, rejecting or excreting all that is superfluous.

Let us examine the matter more closely, however. The further physiological investigation is carried, the more clearly does it appear that the co-ordinated action by which different parts of an organism work together for the maintenance of the whole depends on chemical substances or physical disturbances conveyed from part to part of the living substance. The behaviour of each part is not merely qualitatively, but also quantitatively dependent on these chemical and physical influences. The respiratory centre responds to the minutest alterations in the partial pressure of carbon dioxide in the blood, and the kidney to the smallest variations in the concentrations of water or sodium chloride or hydrogen ions. Everywhere in the living body we find, on close examination, this close and accurate dependence of vital activity on the physical and chemical conditions in the immediate environment. It is, therefore, only from an outside and superficial point of view that there appears to be something in a living organism which acts, within limits, independently of the physical and chemical conditions of environment. The apparent autonomous selective action of the organism turns out to be causally dependent in every detail on physical and chemical conditions. The vital principle, if it exists, is therefore determined in its action by those conditions: we can never isolate and clearly identify its action. As a positive scientific working hypothesis it is thus useless; at the best it only serves to express our ignorance of the exact means by which the parts of a living organism are caused to react in a certain manner to a given physical or chemical change. This being so, the advocates of the mechanistic theory may well ask whether vitalism has any real justification.

To illustrate the force of this question it is only necessary to refer to the main argument for vitalism in the recent writings of Driesch. As is well known he discovered that in the earliest stages of embryonic development the cells of the embryo may be separated completely from one another, or their mutual arrangement may be completely altered by mechanical means, and yet one of the separated cells, or the disarranged collection of cells, may develop in a perfectly normal manner. Here we have presumably disarranged the normal mechanism for development, and yet normal development occurs. From this Driesch concludes that a factor is present which acts independently of physical and chemical stimuli, and which he calls 'entelechy,' and identifies with the old 'vital force.'

Now there is no evidence at all that each cell, in growing and dividing in the one particular manner which constitutes normal development, is not determined by special physical and chemical stimuli peculiar to its position relatively to the other cells, and to the external environment. We do not yet know what these stimuli are; but probably no physiologist would doubt that they exist, and will be discovered when our methods are fine enough. Hence Driesch's argument for an independent vital force breaks down entirely.

A further objection to vitalism is that it

implies a definite breach in the fundamental law of conservation of energy. As already mentioned, every experimental investigation has hitherto resulted in a verification of this law in the case of physiological phenomena. Any 'guidance' of living organisms by the vital principle would imply a creation or destruction of energy; and this would be the case even if the energy created in the living substance were again destroyed before it could escape to the outside, and so become measurable. The reply that this creation or destruction of energy may be extremely small is not one which can satisfy a scientific investigator. A principle which has been verified again and again under all sorts of conditions cannot be set aside except on definite experimental evidence; and this is entirely lacking.

Still another objection to vitalism depends upon the fact that in order to 'guide' effectually the excessively complex physical and chemical processes occurring in living material, and at many different parts of a complex organism the vital principle would apparently require to possess a superhuman knowledge of these processes.   Yet the vital principle is assumed to act unconsciously. The very nature of the vitalistic assumption is thus totally unintelligible.   From this point of view also the hypothesis is useless: for even if we cannot completely understand a living organism by the aid of physics and chemistry, we do not improve matters by postulating an agency which is itself entirely unintelligible.

In one form or another these objections to vitalism have been urged repeatedly, and there has been no satisfactory reply to them. On the other hand, there have been very effective retorts from the vitalistic side against the mechanistic theory.   It is not these retorts that concern us at present, however, but the inherent untenability of the vitalistic position.   If the mechanistic theory is wrong, this does not prove that the theory of the vitalists is right, so that the objections to vitalism lose none of their force.

Let us now try to sum up the position reached at the end of this lecture.   The mechanistic theory of life has been stated in outline.   It evidently affords to biologists **a** perfectly clear working hypothesis—that the peculiar phenomena of life are due to the play of the physical and chemical environment on intra-protoplasmic mechanisms which have been evolved through the influence of natural selection acting for ages. If this theory is correct, the aim of biology is to unravel the mechanism; and biology itself is nothing but a branch of physics and chemistry, which we may, if we like, distinguish as bio-physics and bio-chemistry. The alternative hypothesis of the vitalists has been shown to be unproved, unintelligible, and practically useless as a scientific working hypothesis.

In the next lecture the mechanistic theory itself will be examined.

LECTURE II

CRITICISM OF THE MECHANISTIC THEORY

IN the last lecture it was shown that the position of the vitalists is wholly unsatisfactory; but it does not follow from this that the mechanists are right; and those engaged in the observation of living organisms can hardly escape feeling an instinctive distrust of the mechanistic theory, in spite of the confidence with which it has been urged upon the world during the last fifty years. Somehow or other a living organism never *seems* to be a mechanism, however often it may be called one. The closer the examination, the more confirmed does this impression become, always provided that we are studying living organisms themselves, and not merely their dead bodies, or material which has been removed from their bodies.

Apart from arguments of a general or philosophical character, the main strength of the mechanistic theory arises from the fact that all physiological activity is apparently dependent on physical or chemical 'causes.' Whether the cause be physical disturbance, heat, light, or the presence of one or other of numerous chemical substances, it is always something definite. The physiologist, however, usually calls the cause a 'stimulus,' and it must in the first place be pointed out that in identifying stimulus and response with physical or chemical cause and effect the mechanistic theory makes a gigantic leap in the dark.

To make this matter clear, we may take one or two examples. A minute and scarcely measurable increase in the hydrogen ion concentration of the blood excites the respiratory centre of a normal warm-blooded animal to intense activity. Similar minute alterations in the concentration of water, or sugar, or sodium chloride, or hydrogen ions, have a corresponding influence on the secretory action of the kidney; and cases of a similar kind, where both growth and physiological activity are similarly influenced, may be multiplied indefinitely.

In all these cases the response depends upon the ‘excitability’ of the responding tissue; and when we examine the matter further we find that the excitability is dependent on a very large number of conditions, most of which are presumably still unknown. Moreover, the excitability varies in response to very minute changes in the environment, just as is the case with the original response to the stimulus. The presence or absence of some blood-constituent, normally present perhaps in only the minutest proportions, may profoundly affect the excitability of any tissue : or its response may be modified or ‘inhibited’ by excitation of other parts. It is true that physiologists can obtain a constant response to a given stimulus; but this is only the case if the conditions are ‘normal.’ In living and intact organisms nature provides us with normal conditions. The more carefully the matter is investigated, however, the more clearly does it appear that ‘normal conditions’ designate something of vast and unknown complexity. The fact that such conditions should be realised again and again in different individual organisms, and in the same organism at different

times, is so familiar that its extraordinary
character, from the physical and chemical
standpoint, usually escapes attention.

It may be urged that the immense complexity and delicacy of the physical and chemical
conditions on which physiological responses
depend does not in itself furnish any reasons
for doubting the mechanistic theory of life.
This is doubtless true; but the point must be
emphasised that in the case of stimulus and
response there is in reality no experimental
evidence whatsoever that the process can be
understood as one of physical and chemical causation.   In the case of physiological
stimulus and response no real quantitative
relation can be traced between the supposed
physical or chemical cause, and its effect.
When we attempt to trace a connection we
are lost in an indefinite maze of complex conditions, out of which the response emerges.   It
is of little use to point out that in many cases
the determining cause of a physical or chemical
change may be something very small as com-
pared with its effect.   The turning of a switch,
or opening of a valve, or application of a tiny
spark, may, for instance, produce gigantic

effects. In all these cases we can trace the chain of cause and effect, whereas in the physiological case we cannot. Let it be granted that hope may be entertained of some day tracing the physiological chain. It is nevertheless clear that the existence of such a hope, however confidently expressed, must not be confused with evidence.[1] In recent years it has become the custom to speak of the 'mechanisms' of physiological activities. Until these supposed mechanisms have been actually discovered it would be better not to use language which implies far more than what our present knowledge warrants.

The failure to be able to trace, at present, the causal chain between stimulus and

[1] It appears to me that a very striking instance of this confusion is furnished by the reasoning which runs through Professor Loeb's recently published essay on *The Mechanistic Conception of Life* (Chicago, and Cambridge University Press, 1912). Thus after describing the observation that an unfertilised ovum may be excited to development by various means, such as placing the ovum for a short time in hypertonic sea-water, or simply piercing its outer membrane with a needle, Professor Loeb proceeds to draw the extraordinary conclusion 'that the process of the activation of the egg by the spermatozoon, which twelve years ago was shrouded in complete darkness, is to-day practically reduced to a physico-chemical explanation.'

response implies equal failure in tracing the stream of matter or of energy through a living organism. We cannot tell what exactly becomes of the atoms and molecules which pass into the body—how far, or in what sense, they are built up into the living tissue, or in what way their potential energy is immediately utilised. We must not mistake measurements of the balance of matter or energy entering and leaving the body, for information as to the manner in which this stream passes through the living tissues.

In the previous lecture we saw that in the historical development of physiology there has been a continuous accumulation of facts, obtained by the application of physical and chemical methods to physiological phenomena. The limits which the vitalists have attempted to set to this accumulation of knowledge have been broken down again and again, and there is no reason to suppose that there are any limits. The leaders of the mechanistic physiology have been completely justified to this extent, and we may rest assured that to this extent physiology will never go back on the step which they took. Our knowledge as to physical and chemical conditions which determine excitation and excitability, taking these words in their widest sense as applying to the whole of the elementary phenomena with which physiology and morphology deal, has grown continuously, and will continue to grow. No physiological or morphological phenomenon is beyond investigation on similar lines; but the results of such investigation are by no means identical with the gradual establishment of a definite causal connection between stimulus and response. There has been much unconscious confusion on this point.

We must now proceed to examine the belief, at present very prevalent, that the progress of physiology is leading to a gradual verification of the mechanistic theory of life. It has already been pointed out how dependent the progress of biology, and particularly of physiology, has been on that of physics and chemistry. This undoubted fact has been taken as evidence that physiology is becoming a physical and chemical science, and has commonly been coupled with the assertion that with the help of these methods great progress has been made in resolving physiology into 'bio - physics' and 'bio-chemistry.'

Now it does not follow at all that because physiology makes use of physical and chemical knowledge and methods it must be nothing more than physics and chemistry. All depends upon the nature of the facts revealed by the use of these methods. We might as well argue that because physicists and chemists make use of their sense-organs and brains their science must be a branch of physiology. The real question concerns the nature and general tendency of the con-clusions actually attained by physiologists, by whatever legitimate means these conclusions have been reached.

In surveying the general trend of physio-logical progress it is somewhat difficult to know where to begin. From the early modern times we find the idea present that the activity of some peculiar agency distin-guishes living from non-living things. At the time of Descartes, for instance, it was generally held that apart from the conscious animal spirit or soul, a living body is domin-

ated by specific agencies known as the vital
and vegetative 'spirits.'

Descartes was the first to put forward a
thorough-going mechanistic theory of the
working and development of the animal
body.   His *De Homine* and *De Forma-
tione Foetus* are works of great biological
interest, on account of their influence on
subsequent thought.   The body is represented, in so far as physiological knowledge
extended at the time, as nothing but a piece
of mechanism; and its development is also
represented as a simple mechanical process.
The animal spirits are depicted as a subtle
fluid separated and filtered off into the
ventricles of the brain by a purely mechanical
process.   The afferent nerves are supposed
to be fine fibres leading up to the ventricles
of the brain and connected with valves opening or closing the upper ends of the efferent
nerves.   These latter are supposed to be
tubes conveying the animal spirits from the
ventricles.   As a result of any sensory or
afferent stimulus the threads are pulled, and
the connected valves consequently opened,
with the result that the 'animal spirits' rush

down into the muscles supplied by the efferent nerves, distending them and thus causing them to shorten. According to the arrangement of nerve-fibres or tubes, and valves, one form of reflex action or another follows any given sensory stimulus. The chief motive power for the whole of the mechanisms by which Descartes suggested that the human body is worked and development brought about was furnished by supposed rhythmical expansions of the blood in the heart, owing to chemical explosions.

The greater part of the physiology of Descartes was extremely crude, and lacking in experimental foundation. This crudeness stands out in strong relief against the far sounder, but at the same time far more limited, observations and reasoning of Harvey, whose writings had greatly influenced Descartes. Nevertheless the mechanistic ideas first put forward in a thoroughly systematic form by Descartes have continued to influence biology up to the present day.

If we trace the history of physiology onwards from the time of Descartes we find constant conflict between mechanistic and

non-mechanistic theories, with all sorts of
intermediate shades of opinion. The physical and chemical mechanisms assumed by
Descartes and others who succeeded him
were gradually proved to be non-existent or
non-effective, and new mechanisms took their
place. With each new discovery of structure
or function there came fresh modifications of
mechanistic theories, and fresh objections to
them. The discovery of respiratory exchange
in the latter part of the eighteenth century, the subsequent general application of
chemistry in physiological investigation, and
the introduction of the compound microscope
early last century, were potent factors in this
development.

Up till nearly the middle of last century
the prevailing physiological opinions were on
the whole more or less vitalistic, and they are
fairly reflected in Johannes Müller's famous
text-book of physiology. Physical and
chemical explanations of all processes occurring outside the actual living substance of the
body were freely accepted, for instance, for
all that happens in the liquids enclosed within the body and for its external mechanism

generally. The finer regulation of growth,
nutrition, muscular and nervous functions,
was believed to be regulated by the 'vital
force.' Müller had himself taken a prominent
part in disproving by his microscopical observations on glands the mechanistic theories
previously prevailing with regard to secretion,
and his numerous observations on growth
and development had led him in the same
direction.

Among the younger physiologists about
the middle of the century there arose, however, a very strong reaction against vitalism
and in the direction of mechanistic theories.
Schwann, the discoverer of the fact that the
bodies of the higher animals are made up of
cells, believed that he had also discovered
that cells are deposited from the body liquids
by a process akin to crystallisation. This
hypothesis, of course, struck at the very roots
of vitalism. Within the next few years
attack after attack was aimed at all the
apparent strongholds of vitalism. Du Bois
Reymond published observations pointing to
an electrical theory of propagation of nerve
impulses; Ludwig and others devised new

mechanical theories of secretion and absorption; Liebig, though himself a vitalist, put forward purely chemical theories of various physiological processes; Mayer pointed out the source of the energy of animal movement; and last of all came the publication of Darwin's *Origin of Species.*

Vitalism itself was also vigorously and directly attacked, and its utter weakness pointed out. Perhaps no more clear and convincing demonstration of the untenability of the vitalistic position has ever been given than in du Bois Reymond's introduction to his *Untersuchungen über thierische Elektricitat* published in 1848.

The momentum of the intellectual movement of that time has lasted to the present day, and the influence of this movement has spread in ever-widening circles. But we are not concerned with this, nor with the fate of vitalism, but with the further progress of the experimental verification of the mechanistic theory. From this standpoint all that can be said is that the mechanistic theory has, on the whole, fared very badly. Schwann's simple mechanical theory of growth was based on imperfect observation, and has long been abandoned. We now know that all cells are formed by division of pre-existing cells, and that the problem of the process of cell-growth and cell-nutrition is not one which we have at present any prospect of solving in a mechanistic direction. Nor is it any better with the problems of secretion and absorption. Thanks to the work of Ludwig himself, of Heidenhain, and a host of other investigators, we now know far more than was known at the middle of last century about secretion and absorption ; but their 'mechanism' is further away than ever. When Johannes Müller suggested that the processes of secretion and absorption are akin to the processes of growth he was doubtless far nearer the truth than his immediate successors : for secretion and absorption are evidently only phases of the many-sided metabolic activity which we designate by the name of cell-life. We have made great progress in learning how many-sided, and how orderly, this cell-life is, but none at all in explaining how this life is ordered and maintained.

The simple chemical theories of the respiratory and other metabolic processes occurring in the body have likewise disappeared. The work of Pflüger, Rubner, and others has proved that in the living body all these processes are regulated with the utmost nicety, and that the seat of the regulation lies within the living cells of the body. By what physical or chemical process regulation is brought about we do not know. We have not even discovered the agents employed in the process, although, as already pointed out, there is reason to believe that, in many instances at least, these agents are intra-cellular enzymes.

It has become evident also, that no simple physico-chemical theory of muscular or other physiological movements will suffice. The processes which determine visible movement, or transmission of excitation in living cells, are an integral part of the many - sided activity of the living cell, so that the elementary problems of physiology cannot be solved piecemeal. A physico-chemical explanation of muscular movement, or of secretion, or cell-nutrition, or nervous excitability, would thus be a solution of the whole problem of life. With every year of physiological advance, however, we seem to get further and further away from any prospect of such a solution. It was only through the prevailing ignorance of physiological facts that the physiologists of the middle of last century imagined that they were approaching a physico-chemical solution of elementary physiological problems. To us Schwann's theory of cell-growth seems almost as crude as Descartes' extraordinary theory of the mechanism of growth and development.

In the physiology of the central nervous system the main progress during the last half-century has been in connection with the localisation of function and tracing of paths of physiological connection. Until recently, at least, but little has been done in the direction of a thorough examination of the elementary problems presented by the simplest cases of 'reflex' action. The work of Sherrington and others is now, however, throwing new light on these problems, and it seems quite clear that the old idea of simple and definite 'reflex mechanisms' in the central nervous system must be abandoned. In the nervous system physiologists are also faced with the problem presented by the recoveries of functional activity after destruction of centres or nerve paths on which this activity normally depends. In the case of other parts of the body this recovery of function is also evident enough; but in the central nervous system differentiation of function is so complex and definite that recovery of functions stands out as a fact of extraordinary interest and significance. For this phenomenon it is difficult to imagine any physico-chemical explanation.

To sum up, the application to physiology of new physical and chemical methods and discoveries, and the work of generations of highly-trained investigators, have resulted in a vast increase of physiological knowledge, but have also shown with ever-increasing clearness that physico-chemical explanations of elementary physiological processes are as remote as at any time in the past, and that they seem to physiologists of the present time far more remote than they appeared at the middle of last century.

Probably no competent physiologist will deny this. But, it will be urged, it is only what might be expected that vital mechanism, the product of untold millions of years of natural selection, should be extremely complex ; and the fact that it is so affords no real ground for doubting the mechanistic hypothesis. There are many phenomena in inorganic nature which we do not yet understand, but we do not for that reason regard them as having other than physical and chemical causes.

We must now push our examination a stage further. Let us assume that all the delicate and persistent reactions met with in the living body are due to intra-cellular mechanism produced by natural selection or in any other way, and see whether this assumption will help us to understand the facts. First of all, let us be clear about what such an assumption implies. Those who have not studied the physiology of the higher and better-known organisms, such as man or some mammal, have often very little conception of the exquisite delicacy of physiological reactions. This delicacy was hardly even suspected in former times.

Eustachius, who was probably the first to make experiments on secretion by the kidneys, contented himself with taking a dead, and probably decomposing, kidney, and passing through its blood-vessels such liquids as spirits of wine or water, and observing how much of these liquids issued into the ureter! It is only quite recently that we have come to realise the astounding fineness with which the kidneys, respiratory centre, and other parts regulate the composition of the blood.

To illustrate this point I may perhaps refer to a subject which we have recently been investigating at Oxford. We have found that the respiratory centre is so extremely sensitive to any increase or diminution of the partial pressure of carbon dioxide in the blood that a diminution 0·2 per cent. of an atmosphere, or 1·5 mm. of mercury will cause apnœa, while a corresponding increase will double the breathing. The recent researches of Hasselbalch have afforded experimental evidence of what had already seemed very probable—that the stimulus to which the centre responds is the difference in hydrogen ion concentration, or acidity, brought about by the very slight deficiency or excess of carbon dioxide. He has also investigated quantitatively the effects on the hydrogen ion concentration of the blood of varying the partial pressure of carbon dioxide. From his results and ours it follows that the hydrogen ion concentration of the blood during rest is extraordinarily constant, and remains so day by day and year by year. As the amount of acid and alkali passing into the blood from the food and other sources is constantly varying, it follows that the regulation of hydrogen ion concentration is mainly brought about by the kidneys. It has been known for long that the urine varies in acidity or alkalinity according to the diet ; but Hassel-balch has measured the actual variations in hydrogen ion concentration. Putting to-gether his conclusions and ours, it appears that during ordinary resting conditions the variations in hydrogen ion concentration of the urine are about a hundred thousand times as great as those of the arterial blood.

Thus the kidney epithelial cells react so delicately to variations in hydrogen ion concentration of the blood, that the very smallest variation in the direction of acidity or alkalinity excites them to excrete a liquid which is, *relatively* speaking, intensely acid or alkaline, the net result being that the normal hydrogen ion concentration of the blood remains practically constant.

When we have such figures before us we realise the marvellous fineness of the regulation by the kidneys and respiratory centre. Physiologists are still so much under the influence of the old gross mechanical theories of secretion that attempts at exact measurements of the delicacy of regulation by the kidneys have hitherto scarcely been made in the case of regulation in other directions, though we have every reason to believe that similar delicacy exists as regards the regulation of the water, salts, and other blood constituents. It is hard to realise that something which looks under the microscope like nothing more than a somewhat indefinite collection of gelatinous material can react, and continue throughout life to react, true as the finest mechanism of highly tempered steel, to the minutest change in its environment.

We meet with the same delicacy of adjustment in the regulation of such things as body-temperature, blood-volume, or heat-production. So accurately does the body adjust its consumption of oxygen and of food-material to its energy requirements that it can actually be used, as Rubner found, as a calorimeter; for it substitutes consumption of proteins, carbohydrates, and fats for one another in exact proportion to their energy values as determined with a calorimeter. In Liebig's time this delicate regulation was altogether unknown. It was supposed that any excess of protein simply 'fell a prey' to oxygen, and that if the oxygen percentage of the air was increased, or even if the amount of air breathed was increased, or the barometric pressure was raised, oxidation in the living body would increase, just as in the oxidation of substances outside the body. By accurate measurements of the intake and output of material and energy the true facts as to the rigid accuracy with which metabolism is regulated have gradually been established during the last fifty years.

We must, therefore, in considering the

mechanistic theory, put aside from our minds all the hazy ideas of former generations as to the structure of living cells being nothing but that of an indefinite ' plasma.' Such ideas belong to the early childhood of physiology. We now know that 'simple protoplasm' exists nowhere, not even among the most primitive protozoa or bacteria. Modern investigation of the complex and intensely specific functional activities of every variety of living organism or cell has relegated the old ideas, derived from mere microscopic examination, to oblivion. What the mechanistic theory must assume in the case of an organism such as man is a vast assemblage of the most intricate and delicately adjusted cell-mechanisms, each mechanism being so constituted as to keep itself in working order year after year, and in exact co-ordination with the working of the millions of other cell-mechanisms which make up the whole organism.

This assumption is surely one which taxes scientific imagination to the utmost, but let us make it and continue the argument. We have now to imagine the mechanism of reproduction and heredity. This vast organisation of cell-mechanisms has all been developed from a single cell, itself the product of the union of two cells—an ovum and a spermatozoon; and we have every reason to believe that the hereditary characters, which are derived through both cells, are carried in the two nuclei which unite to form the nucleus of the fertilised germ-cell. On the mechanistic theory this nucleus must carry within its substance a mechanism which by reaction with the environment not only produces the millions of complex and delicately balanced mechanisms which constitute the adult organism, but provides for their orderly arrangement into tissues and organs, and for their orderly development in a certain perfectly specific manner.

The mind recoils from such a stupendous conception; but let us follow the argument further. In the previous lecture I gave such an account of heredity as is usually given from the mechanistic standpoint. It needs very little consideration to see that this account consisted of little more than empty words. The germ-plasm was supposed to be

nothing more than a collection of material of a
certain composition, and capable, in a suitable
environment, of indefinite quantitative increase or growth.   Parent and offspring were
supposed to be similar because they had both
sprung from germ-plasm of the same composition.   This germ-plasm could be added
to, divided, or mixed with other germ-plasm
in sexual reproduction, just as if it were so
much treacle.

It seems perfectly clear that germ-plasm of
so simple a character as this could by itself
furnish no explanation whatever of the development from it of the adult organism with
all its enormous complication and absolute
definiteness of structure.   If the germ-plasm
were so simple, the complication and definiteness would have to be attributed to its
environment : whereas all the evidence points
to the nuclear germ-plasm as the essential
carrier of hereditary characters.   We are thus
compelled, on the mechanistic hypothesis, to
attribute to the germ-plasm, or germinal
nuclear substance, a structure so arranged
that in presence of suitable pabulum and
stimuli it produces the whole of the vast and

definitely ordered assemblage of mechanisms existing in the adult organism. Such a structure must be absolutely definite and inconceivably complex. There is no escape from this assumption; and to speak of 'plasma' when we mean such a structure is clearly absurd.

This nuclear structure or mechanism must, according to the mechanistic theory, have been formed within a very short period by the union of two others—a male and a female one. How two such mechanisms could combine to form one is entirely unintelligible, and the observed details of the process tend only to make it, if possible, more unintelligible. When we trace each nuclear mechanism backwards we find ourselves obliged to admit that it has been formed by division from a pre-existing nuclear mechanism, and this from pre-existing nuclear mechanisms through millions of cell-generations. We are thus forced to the admission that the germ-plasm is not only a structure or mechanism of inconceivable complexity, but that this structure is capable of dividing itself to an absolutely indefinite extent and yet retaining its original structure.

We might, perhaps, get over the difficulty for a generation or two of cells, by assuming that the germ-plasm contains, not one, but several nuclear mechanisms, and that only one of these mechanisms comes into play at each reproduction, others being left to form the germ-plasm of the next generation, or otherwise to provide for emergencies. As a matter of fact, a theory of this kind has been put forward in order to account for the fact that, as has been shown so clearly by Driesch and others, not only the germ-cells, but also other cells in the developing embryo, or even in the adult organism, may, on occasion, give rise to the reproduction of a whole organism. It has been supposed that parts of the original germ - plasm, or, in plain English, nuclear mechanisms capable of giving rise to the reproduction of a whole organism, may and do pass into ordinary somatic cells, as well as into the direct line of future sexual reproductive cells.

This far-fetched hypothesis only makes matters worse for the mechanistic theory of heredity ; for we have to account, not for only one, but for a large number of stupendously complex reproductive mechanisms within the original germ-plasm, and for their endless division and multiplication. The real difficulty for the mechanistic theory is that we are forced, on the one hand, to postulate that the germ-plasm is a mechanism of enormous complexity and definiteness, and, on the other, that this mechanism, in spite of its absolute definiteness and complexity, can divide and combine with other similar mechanisms, and can do so to an absolutely indefinite extent without alteration of its structure. On the one hand we have to postulate absolute definiteness of structure, and on the other absolute indefiniteness.

There is no need to push the analysis further. The mechanistic theory of heredity is not merely unproven, it is impossible. It involves such absurdities that no intelligent person who has thoroughly realised its meaning and implications can continue to hold it.

It may, perhaps, be argued that although a mechanistic theory of reproduction appears to be impossible, this need not affect the practical value of the mechanistic theory in biology. In ultimate analysis the ordinary working conceptions of physics and chemistry present great difficulties ; but no one doubts, for example, the practical value of the hypotheses of mass, energy, atoms, and molecules. Similarly, there may be ultimate difficulties about a mechanistic theory of life, and yet the *practical* value of this theory may remain.

There is certainly some truth in this argument. We often treat a living organism, or some portion of it, as if it were nothing but a collection of ordinary matter, or a machine actuated by the ordinary forces of nature. As will be pointed out more fully below, this mode of treatment is sometimes the best that is practicable, and its value cannot be doubted. In so treating the facts we are, however, leaving out of account almost all those phenomena which are apparently specific to living organisms, and with which biology mainly deals. In actual fact the failure of the mechanistic theory of reproduction cuts deep into our conceptions of almost every detail of biological fact and investigation. Biology deals at every point with phenomena which, when we examine them, can be resolved into metabolic phenomena—exchange of material and energy, as exemplified in growth, development, maintenance, secretion and absorption, respiration, gross movements in response to stimuli, and other excitatory processes. Now metabolism is itself a constant process of breaking down and reproduction of what is living. There is no reason for separating the reproduction of a whole organism from the constant reproduction of parts of it in ordinary metabolic process. Hence our conception of heredity involves every part of biology; and if we cannot frame a mechanistic theory of heredity we are equally at a loss in connection with the ordinary phenomena of metabolism, and we have no right to use mechanistic hypotheses in connection with these phenomena. We have also seen already that the ascertained facts do not in any case point to mechanistic theories of the ordinary activities with which biology deals.

As a physiologist I can see no use for

the hypothesis that life, as a whole, is a mechanical process. This theory does not help me in my work ; and indeed I think it now hinders very seriously the progress of physiology. I should as soon go back to the mythology of our Saxon forefathers as to the mechanistic physiology.

Although the mechanistic theory of life will soon become a matter of past history, there can be no doubt that it has filled an extremely valuable part in the development of physiology. Again and again mechanical theories of one sort or another have served as temporary working hypotheses round which experimental investigation has centred in physiology. This has been as true of the grosser mechanical hypotheses of the seventeenth century as of the more refined physical and chemical hypotheses of later times. The merit of these hypotheses has been that they were capable of either verification or disproof, whereas the vitalistic theories have been incapable of being experimentally tested.

Let us try to see why the latter has been the case. The vital force was conceived as something acting 'from the blue' on ordinary matter, and yet as not acted on itself. Its presence could only be verified through the failure of physical and chemical explanation of certain very striking phenomena, namely the development and maintenance of a certain specific structure and certain specific activities. In so far as any fact connected with life could be explained on physical and chemical grounds the vitalists were at one with the mechanists. In so far as physico-chemical explanation failed they attributed the phenomena to the intervention of the ' vital force.' Thus their dualistic working hypothesis fitted the facts in whatever way they turned out, so that the stimulus afforded by one definite hypothesis, which could be either verified or disproved, was absent. But as they admitted the hypotheses of physics and chemistry with regard to the material and atomic constitution of the universe, and based all their observations and methods on these assumptions, the natural consequence was that in matters of experimental detail they always found themselves dealing either with what appeared to be physical and chemical phenomena or with something unintelligible.

In detail, therefore, the sphere of action of the 'vital force' was a dim and misty sphere of unintelligibility—a purely negative sphere. So far as anything definite could be traced on the confines of this sphere it was something physical or chemical, and beyond this was indefinite mist. Where and how the 'vital force' acted on the atoms or molecules, or what exactly became of them, no one could say; but out of the mist they seemed to appear again, once more in the form of atoms and molecules obeying physical and chemical laws, but ordered in such a specific and unmistakable manner as to indicate that within the sphere of indefinite mist some mysterious factor was at work. As, however, the only definite detail was physical and chemical detail, the mechanists were left in possession of all that could be positively investigated in detail.

The mechanists have contended that the misty sphere is only the mist of our ignorance of the physical and chemical conditions, and that year by year this mist is being gradually dispelled by the advance of physiological investigation. We have seen already that this is a complete illusion. The advance of investigation has only served to make the misty sphere more evident; and not only does it exist, but there is not the remotest chance, as we have just seen, that physical or chemical investigation will ever dispel the mist. The phenomena of life are of such a nature that no physical or chemical explanation of them is remotely conceivable.

Let us try to get to grips with this matter. What is the real cause of the helpless position in which biology finds herself? The mechanistic hypothesis has been the only one of the two which seemed inherently capable of helping us positively in the details of biological investigation; and yet this hypothesis is unmistakably a failure in relation to biological investigation as a whole; and the vitalistic theory, if one can call it a theory, is only a way of registering this failure, and does not help us to a real understanding.

The main outstanding fact is that the mechanistic account of the universe breaks down completely in connection with the phenomena of life. Whether it is not also insufficient in connection with phenomena outside what we at present regard as life is a further question which need not be discussed at present. When any hypothesis fails to correspond with facts it is the hypothesis which needs reconsideration. In the next lecture the physico-chemical hypothesis with regard to our experience generally must first be examined, after which we must consider whether no other hypothesis will fit the facts. It may be that the practical failure of vitalism has depended on the fact that the vitalists have accepted without criticism the physico-chemical account of our experience, and have thus placed themselves in a position in which they are powerless to help biological investigation.

LECTURE III

At the end of the last lecture we were led up
to the question as to how far the interpretation assumed by the physical sciences in their
account of the visible and sensible universe is
valid.   This is evidently a very wide question,
involving far-reaching philosophical discussion.   Yet the reason why such a discussion cannot be shirked is evident.   We found
that in the case of life the facts are inconsistent with the physical and chemical account
of phenomena.   We, therefore, cannot adopt
the attitude that whatever may be the ultimate truth about the universe the ordinary
working hypotheses of physics and chemistry
are sufficient for the immediate purposes of
our work as physiologists.   We must probe
more deeply.

At the present time there is a widespread

prejudice against philosophical or 'metaphysical' discussions. It is asserted, for instance, that system replaces system in the history of philosophy, and no abiding truth is arrived at. The efforts of philosophers are thus vain, and practical men would do well to disregard completely all metaphysical speculation.

Such teaching is unworthy of every tradition which has helped to raise us from the level of primitive savages. Even if it were true that philosophical speculation has hitherto led to no definite result, we should not be men if we gave up the quest after truth. It is only, however, a shallow and ignorant mind that sees in the history of philosophy nothing but a series of systems, each as bad or as good as the other, and succeeding one another like the 'turns' in a music-hall entertainment. The progress of philosophy has been just as continuous as the progress of science, and the history of philosophy appears to be a meaningless succession of systems to those only who have never taken the trouble to understand them.

To philosophers the meaning of that appearance of physical reality with which we are so familiar has been a constant subject of investigation, and we must consider as well as we can what light they have thrown on the particular problem before us.   We must, therefore, try to trace the reasoning which, in its successive developments, has guided them.

When we regard the conception of the visible and tangible universe as it is presented to us by the physical sciences we find that it not only stands the most searching laboratory tests, but that it also stands the tests of practical experience of a very wide kind.   The engineer, the manufacturer, the navigator, the soldier, the lawyer, can apparently rely upon it absolutely.   But philosophy points out that if it corresponds to absolute reality it must be consistent with the *whole* of our experience; and first and foremost it must be consistent with our own conscious relations to it, which have, after all, been entirely left out in the laboratory and other tests.

At first it may seem simple enough to say that we are conscious of the physical world. It is there, plainly before us.   But then comes the reflection that the appearance can only be transmitted to us through our sense organs and sensory nerves. All that we immediately perceive can only be sensory disturbances of some kind, from which we *infer* the existence of the physical reality outside. This is a necessary, but also a fatal admission : for what right have we to draw such an inference? Absolutely none, as Bishop Berkeley first pointed out. We have no right even to call our sensations impressions, or to regard ourselves as anything more than a stream of sensations. The appearance of a sensible world, with our bodies present in it, can be nothing but an appearance due to the manner in which the sensations group themselves. The appearance of substantiality or of cause and effect can be due to nothing else but the mysterious fact that certain sensations are associated together or follow one another in a certain order. Such was the reasoning of David Hume; and his inferences follow inevitably if we start with the provisional assumption that the physical world as science represents it to us has absolute reality. If it has, then we cannot possibly know it ; and all our supposed knowledge is nothing but the illusion which Hume described.

There is no flaw in the reasoning of Berkeley and Hume, absurd as may seem the conclusions which they reached. The flaw is in the premises, and particularly in the assumption from which the reasoning originally started, that the world is something self-existent and outside us, as physical science appears to teach. This assumption simply destroys itself, leaving nothing but the sceptical conclusions of Hume.

How long will it be till the world, and particularly the scientific world, begins to take in the significance of David Hume's reasoning? His body has lain quiet at the foot of the Calton Hill in Edinburgh for nearly a hundred and forty years; but the old ideas which he finally showed to be untenable are still popularly accepted, just as if he had never lived. To those who imagine that the secrets of our existence are likely to be revealed in, say, the latest discoveries in colloid chemistry, I would commend a careful perusal of Hume's *Treatise of Human Nature.*

Whether we take the ordinary popular view, as taught by the theologians, that the soul is a non-material entity situated during life within a material body in a material world, or else adopt the mechanistic theory that there is no soul, but only a series of 'states of consciousness' lighted up somehow or other within a material brain, we cannot escape Hume's destructive criticism. This criticism destroys utterly the assumption of a universe of self-existent things.

But the *appearance* of knowledge of our universe and all that it contains does certainly exist and must be accounted for. The task of accounting for it was taken up by Immanuel Kant, who carried us far beyond Hume.

Kant did not satisfy himself with Hume's account of the data of consciousness as a stream of isolated impressions or sensations, cohering or associating themselves with one another in a manner of which we can give no ultimate account. He proceeded to examine carefully the nature of sensation or perception. One thing which he found was that sensations never do exist in isolation from one another, but that each carries with it a reference to other sensations. A sensation, if it is distinguishable at all, is here and now.

This means that it is given in relation to past, future, and co-existent states of consciousness which are indissolubly united with one another. Moreover, each distinguishable element in experience bears with it its relation to the others in a certain order. Were there no such definite relations of sequence and spatial interconnection there could be no perception or experience at all: the 'hereness' and 'nowness' of each element in experience would be impossible. Hence spatial and temporal relations, causal sequence, substantiality, and the other general ideas by the existence of which our experience is ordered, are all given to us in the simplest elements of experience. The supposed possibility of analysing our perceptions into elements consisting of 'simple' unrelated sensations or 'states of consciousness,' is an illusion. There are no such things as unrelated sensations.

Now this, of course, does not imply that the whole visible universe is given to us as an intelligibly connected system as soon as we open our eyes. It does mean, however, that from the beginning the outlines of such a system are present, however dim and indefinite

the details may be. It means also that between ourselves and the world round us, and
between the various things which we find in
the world, there is no ultimate separateness
of existence such as seems to be assumed in
the ordinary physical conception of the world.
All are parts of one inseparable whole.

This is a conclusion of such stupendous and
far-reaching import that it may need centuries
for the world to take it in and even dimly
realise its implications, and where it leads
us to. Sooner or later, however, it will be
realised that the materialism of the nineteenth
century has been nothing but an insignificant
eddy in the stream of human progress. In
Kant's writings his thought was evidently
trammelled by the difficulty of realising how
great a leap forward he was making. Hume's
scepticism had not completely done its work
in his mind, for he still postulates the existence
of a so-called noumenal world of things-in-
themselves which are the unknowable cause
of the constant newness and variety in our
experience. He also retains the idea of finite
individual minds, each armed, as it were, with
general ideas or ' categories ' which convert

into the orderly system of our experience the impressions caused by the noumenal reality. His immediate successors pointed out that there was no reason left for assuming the existence of things-in-themselves outside of us. These supposed existences are nothing but the ghosts of the world of independently existing matter which Hume had shown to be non-existent. The supposed separately existing finite minds are also not proof against Hume's criticism. We must account otherwise for all the variety and 'contingency' of our universe. Both the external world of things and the spiritual world of persons have their existence, somehow or other, in only one Supreme Existence. In the efforts to show in detail how this is so the philosophical movement initiated by Kant exhausted itself for the time; but we shall have occasion to return later to these efforts.

We must now look somewhat more closely at Kant's account of how the sensible world comes to appear to us as it does, and what bearing his conclusions, and those of his successors, have on the great biological problem which is the main subject of these lectures.

Kant enumerated definitely the 'categories' or general ideas under which he believed that our perceptions are ordered. This list seems very artificial, and is based on the old formal logic, but it includes the ideas of substance, cause and effect, and reciprocal action—the ideas of the physics of Kant's time. He himself, it may be remarked, was a physicist and mathematician of no mean repute. The list limits perception to the perception of a purely physical world, such as the physical sciences described, and he had no special category for living organisms. On Kantian principles, therefore, a living organism can only be perceived as a material structure or mechanism. In this respect, Kant was at one with the mechanistic school of biologists. For him, however, the reason why we must perceive organisms as mechanisms is not because they, in themselves, *are* mechanisms, but because the mind is so constituted that it can only perceive them as mechanisms.

Kant's successor, Hegel, pointed out that his list of categories was incomplete in various directions: also that a special category or categories ought to be added for organic life, as the idea of life is one of the fundamental ideas. There is no reason why a category or general conception of life should not be just as much constitutive of our experience as the category of substance. Here, therefore, we have a possible way out of our difficulties with the mechanistic theory of life. In trying to reduce life to physical and chemical mechanism we are perhaps in some way confusing two different categories. Kant's general philosophical conclusions have in any case thrown a quite new light on our conceptions of the physical world, and have taught us that the validity of these conceptions is of a very different nature from what was previously believed. It may be that just as we cannot base physics on the purely mathematical conceptions of extension, so we cannot base biology on the purely physical conceptions of matter and energy. With these possibilities in mind let us return to a discussion of the facts which biological investigation discloses.

What we have first to ask is whether, as a matter of fact, we habitually use, in dealing with the phenomena of life, a fundamental conception or working hypothesis which is different from the fundamental conceptions of the physical sciences, and cannot be reduced to them. The question, for the moment, is not whether we are justified in using such a conception, but whether we actually do use it. When this question is clearly realised there is, it seems to me, but one answer to it, and that in the affirmative. In dealing with life we not only use a whole series of special terms, but these terms appear to belong to a specific general conception which is never made use of in the physical sciences.

Life manifests itself in two ways—as structure and as activity. But we also recognise—a biologist feels it in his very bones—that this is *living* structure and *living* activity. Each part of the structure not only bears a more or less definite spatial relation to the other parts, but it is actively maintained in that relation. The structure is thus in itself the expression of the activity, and the ceaseless metabolic activity of which visible structure is the sensuous expression forms one department of physiological study—that of nutrition. But the more closely living activity in general is examined, the more clear does it become that all living activity is structural or metabolic activity, either directly or indirectly. The changes in the retina when light falls upon it are metabolic or structural activities. The same is true of the activities of nerve cells, muscle cells, gland cells, or any other living cells; and the gross visible movements of the body, no less than its gross visible structure, are but the outer sign of metabolic activity.

The body can also be affected mechanically or chemically by influences from without; but effects so produced are of not the slightest interest to a biologist, except in so far as they may be connected with living activity.

The living structure is evidently organised : that is to say every part of it bears a definite relation to every other part. As, however, the structure is the outcome of metabolic activity, it follows that the metabolic activity of the living body is also organised, every aspect of it bearing a definite relation to

every other aspect. That this is actually so
has become more and more clear with the
advance of physiology, particularly in recent
times. The fundamental mistake of the
mechanistic physiologists of the middle ot
last century was that they completely failed
to realise this. Such processes as secretion, absorption, growth, nervous excitation,
muscular contraction, were treated as if each
was an isolable physical or chemical process,
instead of being what it is, one side of a
many-sided metabolic activity, of which the
different sides are indissolubly associated.

The relation of the living organism to its
environment is no less peculiar and specific
than the relationship of the internal parts and
activities of the organism itself. Between
organism and environment a constant active
exchange is going on. But this exchange, in
so far as it has any physiological significance,
is always determined in relation to the rest of
the living activity of the organism. Whether
material is to be taken up or given off,
whether and to what extent the organism
is to respond to any 'stimulus,' all this is
determined in relation to the life of the

organism as a whole. The living body and its physiological environment form an organic whole, the parts of which cannot be understood in separation from one another.

Our ordinary language as applied to life corresponds with these characteristics. We naturally speak of a living organism as an autonomous active whole, and think of it as such. The idea of its being a mechanism made up of separable parts, and actuated by external causes, is wholly unnatural to us, and becomes more and more unnatural the more we know about organisms.

The concept we are using is radically different from any physical concept: for in conceiving what is living we do not separate between matter or structure and its activity. The structure itself is conceived as active— as alive.

But the objection may be raised that this is only a loose and inaccurate mode of thinking and expression: for we know that the living substance consists of nothing but matter, though we do not yet know in what exact form the atoms or molecules are combined. This, it must be pointed out clearly, is simply to beg the whole question. It was the complete and hopeless failure of the material conception of a living organism that led to our present inquiry. We cannot admit that the living 'substance' is material. It is the very existence of matter as such that is in question—the adequacy of the concept 'matter' to express the phenomenon we are considering. Let us make no mistake as to what we are really discussing. We have parted company once and for all with the mechanistic philosophy—the notion of a real and self-existent material universe; and we must remember where we now are.

What we have found is that the conception of the living organism is in common and ordinary use, and differs radically from any physical conception. We have also seen that there is no philosophical reason for rejecting this conception. There is no *a priori* reason why we should not, if it helps us, take it as the fundamental conception for biology, just as the physicist takes the conceptions of matter and energy as fundamental for physics.

Before going further we must consider a

preliminary objection.   Whatever the nature
of the actually living parts of an organism
may be, all that we can do in investigating
them, it may be pointed out, is to observe
and measure the physical and chemical
changes resulting from their activity.   We
can measure the shortening of a muscle,
the pull it produces, the oxygen it absorbs,
the electrical changes which accompany
its excitation.   All these are physical and
chemical changes, however ; and the whole
of physiology consists, and can only consist,
of such observations and measurements.   It
must, therefore, from the very nature of its
data, be a physical and chemical science in so
far as it is a science at all.

The reply to this is that apparent physical and chemical changes are the signs or
sensuous data which point to the underlying
living activity.   Just as the physicist has no
direct detailed knowledge of matter, but infers its properties and measures its amount
from various sensuous data, so the physiologist infers the nature and activities of a
living organism from sensuous data.   But to
the physiologist the outward appearances of

physical and chemical change are the sensuous data, by bringing which into relation with his guiding idea he arrives at physiological knowledge; and what he sees behind the appearances of changes in form, electrical changes, absorption of oxygen, and all the other outward signs of muscular activity is the metabolic activity of the living muscle-cells.

If we assume that the conception of the living organism is the fundamental conception of biology, it is clear that the aim of biology differs entirely from what it would be if the mechanistic theory were accepted. All attempts to trace the ultimate mechanism of life must be given up as meaningless. The aim of biology becomes a very different one— to trace in increasing detail, and with increasing clearness, the organic determination which the ground conception postulates. The bodily processes—for instance, the apparent mechanical or chemical processes of movement of the limbs, of breathing, of circulation, of digestive changes, of the taking up and giving off of various forms of matter and energy— become nothing but the expression of organic activity. Their maintenance and working during life are only phases of the organic determination which is the key to all the phenomena of life. They must be looked at from the physiological or biological standpoint, and not merely from that of the physical sciences. The details of bodily structure must likewise be interpreted as the expression of organic determination.

Now it seems to me that the actual progress of physiology, and of biology generally, corresponds exactly with the increasing realisation of this aim. It has already been shown that in tracing the history of physiology we find that on the whole there has been no apparent progress whatsoever towards the mechanistic goal. On the other hand it is perfectly plain that there has been enormous progress, not merely in ascertaining isolated facts, but in tracing organic determination in every detail of bodily structure and physiological activity.

To illustrate this conclusion in detail it would be necessary to present to you, not a single lecture, but a whole text-book of physiology. I shall, however, try to give a single illustration by tracing roughly part of the progress of our knowledge in relation to the regulation of breathing.

From the earliest times it was of course known that in all the higher animals breathing is an essential part of organic activity, as mechanical stoppage of the breathing causes rapid death. Very little further was known, however, until the time of Black, Priestley, and Lavoisier, although Mayow, the fundamental importance of whose work was not at the time appreciated, had in reality come very close to modern ideas. With a single flash the whole subject was illuminated by the definite discovery of oxygen and carbon dioxide and their relation to respiration. Henceforth we knew that the necessity for breathing, and the regulation of breathing, is bound up with the necessity for absorbing oxygen and giving off carbon dioxide. Breathing is of fundamental physiological importance, because the consumption of oxygen and removal of carbon dioxide are fundamental organic activities in animals.

Assuming that an organism is an organism, and not a mere machine, we should expect to find that these activities are organi-

cally determined—determined as in definite
relation to the whole functional and structural
activities of the organism, and not merely
dependent on specific and one-sided conditions,
such as the abundance of oxygen in the in-
spired air, or its freedom from carbon dioxide,
or some mechanical action of the nervous
system, unrelated to the central organic determination.

Misled by mechanistic conceptions of life,
various physiologists have, as a matter of fact,
put forward one-sided theories of this kind,
and they have invariably turned out to be
wrong. As an example we may take the
theory that owing to the structure of the
respiratory centre, and the properties of the
afferent nerves coming to it from the lungs,
breathing goes on, under ordinary conditions,
automatically and without reference to the
removal of carbon dioxide or intake of oxygen.
It was believed that as distention of the lungs
excites fibres in the vagus-nerve which inhibit
inspiration, and collapse of the lung correspondingly excites fibres which excite inspiration,
normal breathing is regulated by this mechanism, though the centre also responds to any

unusual want of oxygen or excess of carbon dioxide in the blood.

Now it seemed extremely improbable that such a theory could be correct; since living organisms do not regulate their affairs in this mechanical manner. Thus guided, we reinvestigated the whole question in the Oxford Physiological Laboratory, and found that as a matter of fact the breathing is, under normal conditions, so regulated in man as to respond with almost incredible exactness to the slightest variation in the output of carbon dioxide or partial pressure of carbon dioxide in the alveolar air. The vagus nerve-fibres play no part in the main regulation, to which their influence is only subsidiary, though not unimportant.

Under ordinary conditions the constancy in the pressure of carbon dioxide in the lung alveoli provides for the supply of oxygen to the lungs and blood; but quite evidently the breathing is, under abnormal conditions, regulated also in direct relation to the oxygen supply; and the action of the lung epithelium, the concentration of hæmoglobin in the blood, and the mode in which oxyhæmoglobin is dissociated in the blood, play an important part in helping this regulation.

The idea which gives unity and coherence to the whole of the physiology of respiration is that of the organic determination of the phenomena. The same idea has to a greater or less extent already given, or is in process of giving, unity and coherence to the phenomena of nutrition, secretion, and circulation. It is an idea which guides us at every turn in physiological work, and constantly suggests new lines of investigation. To leave it out of account in physiology, or to treat it as a mere 'heuristic principle' of very uncertain value, seems to me about as foolish as it would be to reject the idea of mass in chemistry, and retain the phlogiston theory, as Priestley and Cavendish actually did till their deaths. By regarding the structure and activities of a living organism as the expression of organic unity we arm ourselves with a theory which is just as useful in biology as the idea of mass in chemistry. Neither the idea of mass nor that of organism will enable us to predict everything in the chemical and biological worlds respectively, but they both help us enormously in reducing our observations to order.

When we turn from actual physiological investigation to the current text-books on the subject we are confronted by the fact that in accordance with prevalent philosophical beliefs among physiologists, these books are mainly written from what is essentially a mechanistic standpoint, and follow more or less closely the general plan of Ludwig's famous text-book of sixty years ago. An account is given of the physical structure and chemical materials found in the body and its environment, and this account is combined with an analysis of how the whole works when the various elements so distinguished are brought into relation with one another during life.

In this general plan there is absolutely no place left for the living organism as such. We find, moreover, that the various 'mechanisms' to which the headings of the various chapters correspond—the mechanisms of neuro-muscular activity, nutrition, secretion, absorption, etc.—exist, for both reader and author, only as ideals on paper ; and at the end the reader is perhaps inclined to become a vitalist—perhaps not. The author, too, may probably have displayed suspicious vitalistic tendencies. One thing, however, is pretty clear—that the information supplied with regard to the central or 'elementary' problems of physiology is often so vague as to be of little apparent practical importance. The practical physiological information which a student of medicine needs and uses is to a large extent only picked up afterwards at the bedside or in the study of experimental pathology.

What is the reason for this defect in physiological teaching? One might at first suppose that whatever general theoretical opinions might be held by a physiological teacher, yet the facts of the science itself must be the same, and that in teaching the facts, as undoubtedly he does to the best of his ability, he must be doing all that is possible. But here comes in the question, what facts? The facts recorded by physiologists are absolutely indefinite in number, and only the more important ones can be taught. Hence, in accordance with the general plan just alluded to, only those facts which bear on the 'mechanisms' of the various physiological processes can be taught. If, for instance, we are teaching the physiology of the kidneys, we must teach the main facts bearing on the possible mechanism of secretion of urine. We must thus discuss the possible influence of filtration, diffusion, etc., in the process, leaving out of account all details which are irrelevant to this discussion; and when at the end it turns out that the essential mechanism of secretion is quite unknown there is nothing further to do than pass on to the next subject. Actually it is known that, mechanism or no mechanism, the kidney fulfils its functions of regulating the composition of the blood, and that it does so with marvellous delicacy; but facts relating to this do not fit into the plan of exposition of the subject, and have too much of a smack of old-fashioned teleology about them. Hence they are ignored completely, or scarcely touched upon.

It is the same with circulation, respiration, and every other part of physiological knowledge. The fact that the body lives as a whole, each organ or part fulfilling its proper functions and adapting itself to every change, is scarcely touched upon, while a vast mass of unrelated and unassimilable mechanical detail is carefully recorded and described.

The fault evidently lies in the general plan of exposition. This plan does not fit the facts to be described. The living body is a living organism, and physiology must treat it as such, or else submit to the reproach of being a complete failure. The attempt to treat physiology on the principles so clearly laid down in Ludwig's Introduction to his text-book is as wide of the mark as would be an attempt to treat painting and sculpture on the basis of a mere knowledge of the chemistry and physics of paint and marble. Teaching and investigation must begin, continue, and end with the presupposition that the body is a living organism, which must be seen as a whole if it is to be seen at all. To shut our eyes to the central fact of living organic existence is to shut our eyes to physiology itself, and to biology generally. It does not matter what aspect, or what portion of physiology or biology we are studying: we are always face to face with living organisms as wholes or parts of organic wholes. The divisions of the subject are due, either to the particular organisms studied, or to the particular methods of investigation which we happen to be capable of using to advantage. In so far as we are in earnest with the work, and are not blinded by wrong theories, we are always, both in physiological and morphological investigation, studying an organism as a whole.

In the case of the higher organisms we are of course dealing with a compound organism; and we have all shades between highly organised compound organisms and more or less indefinite collective organisms such as a colony or a whole species. In these organisms there is constant active maintenance, constant renewal, constant breaking down and reproduction of the living structure; and this is of the very essence of our conception of life. Reproduction is not in itself a problem, but an axiom; for all living structure is active structure; and it lives in actively maintaining itself and reproducing its structure.

In normal reproduction and death, and in

all instinctive social activity, the individual organism shows itself to be more than a mere individual : it belongs to a wider organic whole. Death is certainly not the mere mechanical wearing out of the organic machine : for the organism, as we have already seen, is no machine. Not only in reproducing itself as a whole, but also in the metabolic processes by which it is changing its substance at every moment, does it flout such a conception. So far as we can at present understand the matter, the physiology of death, and that of reproductive and social activity in all their wide ramifications, belong to the physiology of the species. The individual organism, like the individual cell in a complex organism, belongs to a wider organic whole, apart from which much of its life is unintelligible.

It may be argued that in postulating such a thing as the existence of life as a specific entity we are abandoning all attempt at explanation. We are certainly abandoning the attempt at causal explanation; but in doing so we are only abandoning the philosophical ideas which have already

been shown to be defective and irreconcilable
with experience in relation to the phenomena
of life. This part of our experience implies
the conception of life. By a process of
abstraction from the distinctive facts of life
we can, it is true, regard organisms as simply
so much matter, with so much energy
passing through them. If they had turned
out to be capable of interpretation as
mechanisms this would have been no abstraction, but a correct way of regarding
them. But since they are what they are,
their structure must be regarded as living
structure, and their activity as living activity, both structure and activity being the
expression of an organic and indivisible
whole. The ideas of matter and energy are
nothing but ideas, and in the case of life
these ideas are united and transformed in the
idea of the living organism.

This leads us back to the philosophical
conclusions of Kant and his successors. For
Kant the categories or general conceptions
under which our experience is ordered were
so many separate conceptions unrelated to
one another. We might roughly compare

them to the chemical elements as they
appeared to science before the discovery of
the periodic law and the breaking down and
probable building up of elements. Hegel
pointed out that not only was Kant's list
of categories incomplete, but that the categories bear a natural relation to one another,
the lower being the more abstract or empty
general conceptions furthest from reality, and
the higher being the more concrete and
definite ones nearest to reality. If, for instance, we regarded our experience as consisting of nothing but qualitatively different
data we should be using an empty and
abstract category. If we regarded it as a
world of substances acting on one another
we should be seeing it under a much less
empty category, and the qualities and their
changes would now appear as the qualities
and changes of substances. If we regarded
it as a world of living organisms we should
be using a still higher and less empty
category; for substance, quality, and activity
would now be united in a conception out
of which they all issue, the activity being
no longer regarded as an accident of the

substance or matter, and quality being no longer a mysterious attribute of substance.

It is evident that we cannot apply this or that category at will. Specific categories seem, as it were, to be embodied in all that we experience. By no mere arbitrary process of thought can I make a piece of stone into a living organism or *vice versa*. I can, however, deliberately abstract from what the living organism really is, and regard it as simply a material system. This, indeed, is what the mechanistic theory of life invites me to do. I can also abstract from what the stone is, and regard it as simply a patch of colour. This, or something like this, was what Berkeley and Hume invited us to do, pointing out that the stone, as a substance outside us, is only a metaphysical product of our imagination, and that all that is really experienced is a patch of colour. But, as Kant showed in principle, we cannot thus divest our world of its meaning: we cannot reduce higher to lower categories, and thus explain the higher away: we find that the higher categories are embodied in the very texture of our experience.

When we examine the process of knowledge itself we find that it is a progressive defining of our experience in terms of fundamental conceptions or categories: also a gradual passing from lower, more abstract or indefinite conceptions to higher, more concrete or definite ones. This is the course of all scientific investigation. It is only with infinite travail and pains that our experience gradually defines itself in terms of higher and more definite conceptions. A living organism is not given to us complete in thought all at once: it only gradually reveals itself more and more definitely in the course of long and arduous biological investigation. It is the same on a lower plane for the physical world, or for the mathematical world of abstract form and quantitative relations. But from the very nature of the categories or fundamental conceptions themselves all true knowledge must be a gradual revelation of the lower or more abstract in terms of the higher or more concrete aspects of reality; and as the conception of organism is a higher and more concrete conception than that of matter and energy, science must ultimately aim at gradually interpreting the physical world of matter and energy in terms of the biological conception of organism.   No lower claim than this will satisfy the ideals of biological investigation : of this we may be well assured.

At the time of Kant and his immediate successors biology had hardly begun to be conscious of her strength.   Living organisms seemed, as it were, to be at the best only dotted about here and there in the midst of a totally foreign physical universe.   Kant assigned to them a very doubtful place in his philosophy, and Hegel in his *Philosophy of Nature* represented Nature as a sort of waste in which any and every kind of existence is strewed about indiscriminately—a conception repugnant to true science.   Meanwhile the advance of the natural sciences has profoundly changed our outlook on Nature. First and foremost we have come to realise the fact of evolution.   The true significance of this fact is a very different one from what is still very generally supposed.   At first evolution appeared in popular thought, embodied, for instance, in the writings of Herbert Spencer, as a derivation of the organic from the inorganic. We have seen already, however, that, evolution or no evolution, there is not the remotest possibility of deriving the organic from the inorganic. Evolution, therefore, takes on a very different significance. In tracing life back and back towards what appears at first to be the inorganic we are not seeking to reduce the organic to the inorganic, but the inorganic to the organic. The apparently indefinite microscopical aggregations of formless colloid material—'sarcode' or 'protoplasm'—which were at first taken for the origins of life from the inorganic, have gradually turned out to be definite living organisms. But biology will not stop at these: for they must have been evolved from something more primitive. She will gradually push her advance victoriously further and further into the domain of the apparently inorganic.

The premonitory signs of this advance are not lacking. For nearly a century—since the days of John Dalton—the chemical atom was the ultimate term of physical and chemical investigation. The world seemed to be resolved into a world of atoms, in which, as du Bois Reymond pointed out, there was no place for life in any such sense as that which I have endeavoured to depict in this lecture. Recent discoveries in physics and chemistry have, however, completely shattered this conception, and with it that of matter and energy. We now see physicists and chemists groping after biological ideas. No one can yet tell what conceptions will emerge from the ruins of the atomic theory; but it is at least evident that the extension of biological conceptions to the whole of Nature may be much nearer than seemed conceivable even a few years ago. When the day of that extension comes the physical and chemical world as we now conceive it—the world of atoms and energy—will be recognised as nothing but an appearance, though for practical purposes it will still remain very useful. It will stand fully confessed as a world of abstractions like that of the pure mathematicians. Meanwhile it is already a world of abstractions to the biologist who has faith in the principle of evolution and also in the fundamental conception of the living organism.

To one point with reference to the biological conception of a living organism no reference has yet been made. Living organisms always, or nearly always, appear to be marvellously 'adapted' to their physical environment. Yet this adaptation is not of the essence of the biological conception of an organism. We know, also, that organisms may develop which, in one way or another, are so misshapen or defective that they cannot survive, though they have all the essential characteristics of organisms. They maintain their existences as organisms for a short time, blindly struggling, as it were, to preserve the defects which make them incapable of surviving.

Organic life is 'blind': each organism blindly struggles to maintain its own existence or that of its species. We can only explain the actual marvellous and intricate 'teleological' adaptation of organism and physical environment by the fact that, as pointed out by Darwin and Wallace, those organisms which are not so adapted must disappear. Even if we could see far enough to be able to regard as organic the whole of what appears as inorganic, the entire world would still appear as a blind struggle between different organisms. Of this conflict, whether we regard it as conflict between organisms themselves, or between organisms and physical environment, we can give no biological interpretation, and are thus forced back on a physical interpretation. From the standpoint of ideal biology organism and environment would be one. The organism would be not something in a more or less foreign environment, but in its own environment, which it has grown in, and which has been part of its very nature from the outset. Of the actual foreignness or imperfection in the environment biology as such can give no account.

This discussion brings us to the subject of the last lecture. In the higher organisms, at least, we find distinct evidence of a quite new factor—consciousness. A conscious organism is, as it were, fighting the inorganic world, not blindly but with the weapons of that inorganic world itself. It answers blow with counter-blow, and physical force with counter-force. The conscious organism is aware of the inorganic world as such, and reacts as in presence of that world, whether it be real or unreal.

In concluding this lecture let us survey the progress made in our discussion. We have seen that the idea of the physical universe as a world of self-existent matter and energy is only a temporary working hypothesis by means of which we are able to introduce a certain amount of order and coherence into a large part of our experience. The fact that, as shown in the last lecture, this hypothesis breaks down in connection with the phenomena of life need not, therefore, puzzle us. The phenomena of life involve another and radically different conception of reality, and I have endeavoured to define this conception, and point out that it is actually used as a working hypothesis by biologists, and that by its means we introduce order and intelligibility into biology, whereas there is no such order or intelligibility if the mechanistic theory of life be adopted. The idea of life is nearer to reality than the ideas of matter and energy, and therefore the presupposition of ideal biology is that inorganic can ultimately be resolved into organic phenomena, and that the physical world is thus only the appearance of a deeper reality which is as yet hidden from our distinct vision, and can only be seen dimly with the eye of scientific faith. We have seen, finally, that the ideal biological world is never completely realised.

LECTURE IV

IN the previous lectures I have endeavoured
to follow out what is involved in our observations of living organisms looked at apart from
their conscious activities, and from a purely
biological standpoint. When, however, we
examine conscious activity or personality we
find that we are in presence of an order of
observations differing greatly from those with
which biology deals.

We have seen that a biological phenomenon
apart from the other phenomena, co-existent
in space, which we designate as the life of an
organism, has no definite meaning. In conscious activity, however, there is in addition
no meaning in any phenomenon of perception
or volition isolated *in time* from other
phenomena. A perception includes not only
co-existent perceptions, but also preceding

and succeeding events. In perceiving a printed word on this paper we perceive it not only as part of what else we see at the moment on the paper, but as part of the words seen previously and anticipated later on the same and other pages, and as part, also, of a multitude of other past and anticipated phenomena in our conscious existence. Even as regards the letters of each word the same holds good. The significance of each letter depends on its connection with preceding and succeeding letters. We can see this point still more simply and clearly when we analyse the perception of music. Thus both the past and future, not merely the present, enter into perception. The perception includes the past and the future. The past perceptions are not done with except indirectly through the present effects which they may have produced: they are still directly active in the present; and future perceptions are likewise active in the present, since anticipation is just as important as memory in determining the nature of a present perception. Interpretation, tradition, and ideals are essential elements in perception. Past, present, and future are thus not outside one another in perception, but co-equally present: hence personality is not existence in time, but succession is a distinction or principle of order within personality, as Kant was in modern times the first to point out clearly. The ideal world-equation of Laplace could not be applied to conscious activity, since the separate terms in the equation would be indefinable. The controversy over free-will and determinism has likewise no meaning.

Perceptions belong to an orderly world of experience in which past, present, and future times are co-present. The principles which we distinguish in this orderliness are those of the various mathematical, physical, biological, and humanistic sciences. Degree in the clearness of perception depends on the degree to which these principles are explicitly embodied in perception.

Explicit perception or 'exact' knowledge is quite evidently always selective. One particular principle of order in experience is emphasised to the exclusion of others, so that knowledge represents an abstraction from actual experience. Thus we can distinguish elements in experience as simply arranged outside one another in order of time. This represents the highest degree of abstraction, since the elements of experience are only regarded as arranged outside one another, apart altogether from the various other relationships which also define them, and apart even from the fact that succession is a relation, so that unrelated elements of experience succeeding one another in time would represent a mere contradiction in terms. We can also regard the elements of experience as simply arranged spatially, and this is the attitude of pure geometrical science. In pure kinematics co-present relations of both space and time are alone considered.

The physical conception of experience represents a less abstract treatment of it. Relations in space and time are explicitly recognised as relations of units of something enduring—substance—acting on one another in accordance with definable laws. Hence we can quantitatively distinguish the substance in spite of its modifications in consequence of the mutual action, and also distinguish quantitatively the modifications resulting

from mutual action.   If we neglect the fundamental fact that the physical world is a world
of experience—that substances or 'bodies'
and their reactions with one another have no
intelligible meaning except as related to one
another within our experience—the physical
world appears as a world of self-existent
things, among which our own bodies are
comprised.

The biological conception of experience
represents a still less abstract treatment of it.
The parts of an organism and its environment,
together with their mutual influences on one
another, are recognised as expressions of a
whole which actively maintains itself through
change, constantly reproducing what it loses,
and in development or degeneration still
maintaining the expression of the whole.
Biological experience cannot be stated in
terms of the independent and self-existent
'bodies' of physical experience and their
mutual reactions.   The parts which participate
in life are seen not to be independent of one
another, but only to be definable through
their participation in life.   The behaviour of
the dead body or a dead part of an organism

can be interpreted physically; but between what we interpret as life and what we interpret as mere mechanism there is a gulf across which we do not at present see clearly, as was pointed out in detail in the previous two lectures. We certainly cannot reduce life to mechanism; but we do not yet see clearly how to extend biological conceptions to what we must still distinguish as the inorganic world, or even to innumerable details of what we observe within the living bodies of organisms.

The humanistic sciences deal with the world as a world of conscious experience and conscious activity, and embody a much less abstract and more penetrating interpretation of experience than even that of biology. Experience is regarded as made up of elements which are manifestations of conscious activity or personality. The interpretations which are made use of in the mathematical, physical, and biological sciences are interpretations of human experience and lie within it. In actual fact, for instance, even the simplest succession in time has no meaning except as a revelation of elements co-present to one another in consciousness. Nor have substance or energy or living organisms any meaning apart from the co-presence in consciousness of the successive signs through which they manifest themselves. We can assign no meaning whatsoever to 'objective' worlds of time, space, matter, energy, and life, apart from perceptions of them. David Hume's reasoning on this point is conclusive. The world we know is the world which appears to us within consciousness, and with its elements co-present to one another and exhibiting definite and orderly relationships of succession, extension, identity of substance and energy, life. Hence the humanistic sciences include within themselves the worlds of human abstraction with which the more abstract sciences deal. The latter sciences represent the work of human abstraction, employed for human convenience, and constituting only preliminary knowledge. Their development can be traced in human history; and when, through absence of philosophical understanding, their human origin ceases to be recognised, confusion is inevitable. A science divorced from its historical development, and taught dogmatically as if it represented some 'objective' reality apart altogether from human experience and the long process of human effort which has built up the science, is the dead science of an inferior text-book. Nothing is more misleading than the idea that an uninstructed person has only to open his eyes and passively 'observe' in order to be convinced of 'scientific facts.'

We are apt to regard perception as something passive and entirely distinct from volition or voluntary action. When, however, we look more closely it becomes evident that perception involves deliberate activity in the selection of what is perceived, or in the abstraction of what is allowed to become prominent in consciousness from what is disregarded. A physicist and a biologist placed in a given position will be apt to see quite different things, or interpret them quite differently; and a man with other interests will see differently from either of them. The further action of each of them will accord with their different perceptions. Perception is thus nothing but a step in voluntary action, and indeed the most important step. The further step consisting in overt movements of the limbs, etc., is in no sense more active than perception itself, so that in analysing the nature of perception we are also analysing that of volition.

Let us now examine in further detail the phenomena of conscious activity. When we look at the living body from the physical standpoint we can trace towards the brain impulses transmitted from outside through the sense-organs. Unless these impulses reach the brain through nerves there is no sensation or perception. Moreover if the brain is seriously injured or destroyed all the signs of consciousness disappear. Local injuries of the brain will also result in the definite blotting out of certain kinds of perception, memory, or capacities of voluntary response, so that perceptions, memories, and volitions seem to be localised in different parts of the brain. Thus from the abstract physical standpoint we are apparently driven to conclude that different parts of the brain are the 'seats' of different forms of conscious activity, and that consciousness itself is only an accompaniment, however mysterious, of the physical and chemical changes occurring in the brain.

But now let us look at the facts more closely, as a physiologist looks at them. When

we examine the physiological activities of the
brain we find that it is in constant active
connection through afferent nerves with all
parts of the body and so with the surrounding
environment.   A response to an afferent
impulse through a sense-organ is determined,
not merely by the nature and strength of this
impulse, but also by the coincident inpouring
of impulses from all parts of the body and
surrounding environment. If, for instance, we
cut off a large part of these impulses by section
of afferent nerves—for instance the afferent or
sensory roots of the spinal cord—the result is
paralysis of the parts thus deprived of local
afferent stimuli, although the afferent or motor
nerves are intact.   This paralysis may be as
effective as if the motor nerves themselves had
been severed.   A nervous response is also
directly dependent on the amount and quality
of the blood-supply to the nervous system,
and thus indirectly on the functional activity
of the rest of the organism, and on its relations
to the environment through nutrition and
respiration.   A very slight excess of carbonic
acid or deficiency of oxygen has a marked
effect on the nervous responses to afferent

stimuli; and immediate loss of consciousness results from great excess or deficiency. Equally marked effects result from all sorts of minute alterations in the quality of the blood, such, for instance, as are caused by disease or excision of the thyroid gland or failure of normal function in the kidneys or liver. Thus the response to a sensory stimulus is due also to innumerable inpouring chemical stimuli.

With every advance in physiology the experimental evidence shows more and more clearly that we cannot separate off and specify the occurrences in separate parts of the body, and particularly of the central nervous system, as we can for practical purposes separate off and specify the occurrences in different parts of a machine. In nervous responses, and more particularly conscious responses, the whole nervous system, and mediately the whole organism and its environment, are involved. The response is a manifestation of the whole life of the organism, and not merely the response of the brain or a definite part of it. From the physical standpoint we seem to lose ourselves in an indefinite tangle of reactions when we seek to analyse what is happening in a nervous response. The ideas which Descartes formed of the mechanical working of the nervous system and of the body as a whole were very far from being an adequate representation of the facts. Between his conception and that reached through experimental investigation in recent times there is a profound difference.

When we consider the character of physiological responses as a whole we find that they are of such a nature as to contribute to the maintenance of the normal life of the organism as a whole; and in nervous responses there is no exception to the rule that life is continuous organic maintenance through adaptation to changing conditions. From this point of view the responses become intelligible and therefore predictable, since the undefinable complexity of what we endeavour to interpret as physical and chemical reaction resolves itself naturally just as in the case of other physiological responses referred to in the previous lecture.

An organism considered simply as such is blind to its experience. Its reactions are simply the expression of its nature. It may be said to possess 'organic memory' in the sense that it behaves as if it always remembered its own particular nature and physiological environment as an organism, blindly maintaining and reproducing this from generation to generation. Organic memory in this sense is, however, a different thing from conscious memory. \A plant possesses 'organic memory,' but this does not constitute what can properly be called memory, nor do we regard it as evidence of consciousness. To 'organic memory' we can attribute the acquired ease and efficiency with which some physiological activities come to be carried out —for instance the ease with which the nervous reactions involved in walking, running, or riding a bicycle are made after practice, or the ease with which some specific infection is repelled after a previous attack or the action of a suitable vaccine, or the fact that organic structures tend to grow with use. All such cases seem, however, to be nothing but examples of the same blind processes of active maintenance through adaptation which are characteristic of all life. These processes often require considerable time to complete themselves or to pass away, and they involve not

merely the activities of structure, but structure
itself, since organic structure is an expression
of organic activity. To an altered environment, in whatever way it may arise, what we
interpret as a mere living organism responds
blindly by such modification of activity and
structure as tends for the moment to maintain
the normal. Is this all, however? In the
phenomena of conscious activity we find
definitely that the interpretation just given is
inadequate.

The characteristic distinction of a conscious
organism is that it 'learns' from experience.
This signifies that a conscious nervous response
is determined not merely by the simultaneously
inflowing influences from all parts of the
living body and its environment, but also by
what, in order of time, has preceded or will
succeed. From the physical, or even the
biological, point of view such a fact appears
as a completely unexpected and unintelligible
'revelation.' Yet it is a fact which must be
insisted on as fundamental in the interpretation of our experience of personality. We
are unable to reconcile it with the physical
or biological interpretations; but it is there

as a fact of experience, and if it is inconsistent with the physical and biological interpretations we can only conclude that they are both of them only provisional and imperfect interpretations.

Dynamic memory and prescience—the traditions and ideals of conscious personality—can be distinguished from the mere power of calling up images of the past, or transferring them to imaginary places and times. These images of the past seem to be only the effects of past events—records which can be read, but not in themselves exercising any appreciable direct influence on the present or future. Powers of detailed memory are notoriously different from powers of intelligence and imagination. It is the present and future significance of what is remembered — the embodiment of experience in character—that counts in conscious activity. Detailed memory may be largely blotted out during lapse of time, or by injury to the brain, just as the information which comes through one or more of the senses may be interfered with, or ceases during sleep. The faculty of recording images of events in memory may also be greatly diminished, as when a man is under the influence of considerable shortage of oxygen. Failure of detailed memory is, however, a very different thing from failure of intelligence ; and, indeed, persons are often so overburdened with what they have to remember or observe that their intelligence suffers. A student, artist, or person in any specially responsible position, must have leisure to think. History consists not simply in records but in their interpretation. This interpretation is never complete, since the present and past are in dynamic relation, so that the past is constantly showing its own nature in the present and thus being constantly re-born, while the present is also showing its own nature through the ever-present influence in it of the past. History must thus be read backwards as well as forwards, and is constantly being re-interpreted. Similarly we can trace the development of an individual person through the abiding and ever-broadening meaning and practical outcome of the events in his life-history. It is not the mere isolated events in his history that matter, but the living spiritual unity manifested in them.

Unity of the past, present, and future, and not merely the unity throughout space-separation of a living organism and its environment, is what constitutes personality or spiritual unity. Just as a conscious response is from the physiological standpoint no mere response of the brain but a manifestation of the whole life of an organism, so from the psychological standpoint it is no mere organic response isolated in time, but the manifestation of a spiritual unity which holds time-relations within itself. A person is no mere physical body among other bodies, no mere living organism, but a spiritual being which neither physical nor biological conceptions are capable of representing.

When we look further at the spiritual unity of personality we seem to find that it is centred in an individual self corresponding to what from the biological standpoint is an individual organism. The details of structure and activity in which the life of this organism manifests itself are, however, re-interpreted in personality as 'interests' of value, and the person appears as their meeting-point in space and time, from which time-determinations extend backwards and forwards, space-determinations extend outwards, and interests and values are centred. What I perceive is what interests *me*, and my actions are determined in my own interests.

At first sight it might, perhaps, appear that the surrounding inorganic world is, or at least can be, perceived merely as a physical world of matter and energy, or that if we seem to perceive it as anything else this is mere anthropocentric illusion. It is true that in practical life we regard the inorganic world as a world of our own interests. The engineer or the workman treats the physical world from the utilitarian standpoint of its uses and hindrances to human interests. But reality, we are accustomed to think, is something apart from this accidental view of it, and what we really see is a world of matter, energy, and their distribution as described by the physical sciences.

If we look back, however, at the historical development of scientific ideas we find that they have sprung out of the needs of practical life. The tradesmen who devised and used the balance, and the engineers who measured in horse-power and foot-pounds, had practical ideas of mass and energy long before men of science definitely formulated and extended these conceptions. Physiology originated in medicine, and chemistry in practical arts of manufacture. The sciences are built on ideas which have their roots in human needs. These ideas are changing and developing with human progress, and never escape from being anthropocentric. In the extraordinarily interesting developments of physics and chemistry during the last few years we have been witnessing profound changes in what to the last generation appeared to be conceptions of atoms, mass, time, and space, that had been established for all time. When separated from the mother earth of human experience and dogmatically accepted as ultimate representations of reality, scientific ideas cease to live. What we see in our world is determined by our attention, which itself is determined by our needs. Hence the teleological determination of personality includes the whole of our world of experience, and is what gives unity and coherence to it. Nature as we see her is not a world in which conceptions or 'categories' are strewed about indiscriminately. Kant's 'synthetic unity of apperception' is more than a sort of general enclosure in which this strewing about occurs. In 'Nature' personality is actively present everywhere.

We can in thought abstract from, or leave out of account for the moment, the teleological or humanistic nature of our world of experience, and treat the world as if it, or part of it, existed apart from ourselves. We are in fact constantly making this abstraction, leaving out of account the fact that all our practical and scientific ideas are human ideas, the product of human development, and the outcome of human needs. We can also regard human activity as simply the action of one self-existent thing on other self-existent things outside it. This abstraction from reality is, like all rule-of-thumb methods, useful and even unavoidable in practical life; but it *is* an abstraction from reality; and in considering the nature of personality we cannot make any such abstraction. The self-existent things are in reality the creation of our own thought —the product of logical tools used for our own purposes. The imperfection of these tools and their products is our own imperfection : for they are part of ourselves and belong to our personality. In scientific work we are not simply studying passively something which is apart from and outside us : we are just as much personally responsible for our work and its outcome as any ordinary workman. He has to study his materials and tools just as we do, and our study of them is only part of the teleological ordering of our world, just as in his case. Our theories are practical theories, just as his are, though he is putting them to immediate use, while we are looking to their future use.

The world we see is a world of personality. We can reach this conclusion either from the purely philosophical side, with Hume, Kant, and the other philosophers, through the abstract demonstration that our world is the world as we see it. Or we can reach it from the side of human history, through the more concrete demonstration that the world we see is the world which by painful human effort we have gradually fashioned, and are continuously fashioning further in thought and action. The claim sometimes made on behalf of natural science, that its interpretations represent absolute reality, independent of man himself, cannot be maintained.

At first sight it might seem that the conception of the world as a world of personality involves the inference that the existence of the world is bound up with our own individual existence—with the existence of a person who was born and will certainly die. Such an inference would, however, be wholly mistaken. There is nothing more certain than the existence and compelling power of duty which is no mere duty to the individual self, and of truth which is no mere truth for an individual. This signifies at once that personality is no mere individual personality. The personality of any individual represents a spiritual inheritance and environment living and active throughout his present, past, and future. He sees with the eyes of his mother, father, teachers, and numberless others with whom he has come into contact: also through the light of oral tradition, books, and other records. His perceptions are equally the expression of inherited character already showing itself in his earliest infancy. He is what he is through his relations to others, and particularly to the stock he has sprung from. His personality was no *tabula rasa* at birth, nor as already shown could the metaphor of markings on a tablet represent his perceptions. He is his spiritual self—a self not existing in time or space, but within which distinctions of time and space, of matter, organisms and other persons are present. The other persons are not outside his own spiritual self, but within it. He feels, perceives, and acts with and through them, just as he feels, perceives and acts through his own sense-organs, brain, and other organs. Individuals as such are born and die, but the personality which is their reality cannot be born or die.

This may seem to be mere picturesque imagery or unintelligible mysticism, but is not so. It represents the conclusion which philosophy leads up to; and philosophy is only the analysis of experience, taking the whole of experience into account, and not merely certain aspects of it.

Biological investigation has shown that life is continuous from generation to generation. Pasteur's investigations put an end finally to the old beliefs in spontaneous generation. There is no break between parent and offspring, between individuals of the same species, genus, or family; and though as yet we cannot clearly trace life back into what we at present, for want of fuller knowledge, call the inorganic world, this may be possible in the future. We also know that the life of an individual higher organism is the expression of the lives of the countless cells which we can distinguish in its body and which are constantly dying and being born: also that the life of a cell is itself the expression of the lives of countless simpler units of life, of which part are always disappearing, while others reproduce themselves. The individual unit of life reveals its true nature in the wider life which it participates in, and this nature is revealed in dying just as much as in living.

It is no otherwise with personality, although personality cannot, like life, be regarded as an existence within time. It is only in social life that a person realises his real nature. Regard for truth that is truth for others, and regard for duty that is duty to others, belong to his nature. Personality must also be continuous from generation to generation, though for the present we seem to lose clear sight of it in early infancy, and completely lose sight of it in germ-cells and other individual cells or lower organisms. We can, however, trace its continuity in human history and ideals; and we realise its existence through the answers to the awakening calls of what appeals to humanity, or even to a race of men. The prophets are those who make calls that reveal past, present, and future.

During the British retreat from Mons in August 1914 a rumour arose that there had been seen among the rear-guards the forms of old-time English bowmen firing their arrows at the advancing enemy. How this rumour originated no man could tell, but in spite of its apparent absurdity it put heart into the retreating army, for there was deep vision in it. The bowmen were there, but dressed in khaki, and fighting, not with bows but with rifles, machine-guns, and field-guns. Side by side with the bowmen, and also dressed in khaki, were the stout Scottish spearmen who had faced them at Bannockburn and Flodden. But those advancing men in field-grey were

also men of old history. Each had answered
his country's call; and had it not been for
such men on both sides, and not merely on
one side, the war would have been ended in a
few days, in spite of all the accumulated
armaments and organisation. From generation to generation there can be no break in
personality.

As was remarked in the previous lecture, we
seem unable, from the purely biological standpoint, to give any account of progressive evolution except as the outcome of a blind struggle
for existence. But for conscious personality
the struggle is no longer blind : the future is
foreseen and fore-ordained if only to a limited
extent; and the past is remembered and
acted on. This is not only so for individual
persons but the traditions and ideals of a race
represent its memory and foresight. From
the standpoint of personality evolution takes
on a new aspect, and is no longer a blind
process.

In human beings we cannot as yet distinguish
clearly between the conscious existence of the
individual as a whole, and that of sub-person-
alities, although we seem to come near this in

the case of dreams and various abnormal
mental conditions. Psychological study will
certainly bring us still nearer. We may confidently anticipate that just as individual
members can be distinguished in a society,
so also distinguishable elements of personality
will be discovered in what we ordinarily call
individual persons. This does not mean, however, that these elements are self-existent;
for it is in the whole individual that they
manifest their individual being. Similarly it
is only in the wider personality revealed in
social existence that the individual members
of it manifest their true being. They have
sprung from this wider personality, are born
in it, and exist in it; and unless they
realise this they do not realise what they
are. Their wider personality is no ghost-
like abstraction, but just the ordinary, but
infinitely wonderful, spiritual reality which
we see and feel around us everywhere in
time as well as in space.

From the standpoint of personality the past
and future are not outside the present, nor
the parts outside one another in space: yet
the meaning of personality is only realised

and expressed in space and time relations. However wide the conception of personality may be there is always a fleeting ' here and now' in which it appears to be centred, although from its very nature it cannot be so centred, since time and space are within it. With the here and now there is always the appearance of indefiniteness, imperfection and ignorance. We may see that the universe can only be the manifestation of timeless personality : yet we cannot see in detail how this is so. It appears to us as full of mystery, imperfection, suffering, and endless contingency ; and yet over and through all these appearances personality enfolds within itself time, space, and all apparent contingency and imperfection.

We can find a clue to this mystery if we look back at the distinction between mechanism and life, and consider how the conception of life is related to that of mechanism. It is only in the failure of mechanical conception that we reach the biological conception. But in this failure the mechanical aspect of Nature does not disappear. If it did there would be no meaning in the biological aspect. If in respiration oxygen did not appear to us, and continue to appear, physically and chemically as what we call oxygen, and the structures involved in respiration did not appear to us as physical structures, we could form no conception of respiration and the structures involved in it as being manifestations of life. It is true that when the phenomena are looked at as a whole we see life; but life only manifests itself through control of what appears as physical and chemical disorder—through the appearance of constant active struggle. An ideal otiose life, in which everything went rightly without effort, would not be life at all. There could be no life without the appearance of mechanism, just as there could be no whole without the appearance of self-existent parts. Biology implies physics and chemistry, and grows in meaning and richness with their growth, although, as we have seen, it cannot be reduced to physics and chemistry. This would still be so if we could see the way to extend biological conceptions to the whole of the so-called inorganic world. Physics and chemistry would still exist as indispensable abstract sciences.

The truth, therefore, is that the lower and more abstract aspects of reality are implied in the higher aspects, and do not disappear in them though constantly disappearing. Both lower and higher aspects are constantly growing —constantly being born afresh. The conception of mere life, with its existence in time, is constantly disappearing in that of personality, but never disappears. A perfect universe of personality, with no imperfection and suffering, no here and now, no progressive science, no progressive ordering, in thought and deed, of the past, present and future, would not be an actual world. To use a mode of expression employed by Professor Pringle–Pattison in his recently published *Gifford Lectures*, the lower aspects are ' organic ' to personality.

Through analysis of what experience involves we are led up to the conception of the Universe as personality. In our relations to fellow-men, fellow-animals, and Nature as a whole, we find that this personality is not that of an individual man, but that all-embracing personality which we call God. This conclusion can be drawn, though less directly, from the mere study of Nature : for not only is such knowledge of Nature as we possess the expression of human personality, but we are the children of Nature. This means, not that we are mere matter or mere living organisms, but that Nature, too, since we are of her, must be of personality. This is the real outcome of a thoroughgoing application of the doctrine of evolution. So far as our present direct scientific knowledge goes, the development of higher from lower forms of life appears as the outcome of chance conditions of environment acting on chance variations among organisms. But what lies hidden behind the words ' chance ' and ' variation ' ? In conscious activity the variations which represent the progress of civilisation, or of education in the individual, are not the outcome of blind chance, but are expressions of personality to which both future and past are present, however imperfectly. We have thus no right to attribute to chance the variations which occur in lower organisms in which we cannot yet trace directly the evidences of consciousness. Nor have we any right to attribute to chance that which we describe as the physical and chemical constitution of the inorganic world. That something more than chance determines this has recently been pointed out by Professor L. J. Henderson in his books *The Fitness of the Environment,* and *The Order of Nature.*

Since personality embraces all experience there is nothing outside it. In one sense, therefore, personality is always complete. In another sense, however, it is always incomplete, since it is only through the constant negation of mere appearance that personality realises itself. Hence mere appearance is always present and constantly rising up within us as the here and now, the uncertainty, obscurity, imperfection, sin, and sorrow of the world. The distinctive note of Christianity is that God is present in this imperfect world, and that through the sin and suffering, and the constant negation of it, oneness with God is reached. This is the Gospel which Jesus and his followers preached.

Let us now look back at the course of the discussions running through these lectures. The first lecture stated in outline the physico-chemical or mechanistic account of life. It is quite evident that this account, so far as it goes, is of great practical value, since it enables us to predict much of what we observe in the behaviour of living organisms. It also seems at first sight to help us towards a general philosophy of what we call Nature.

In the second lecture the mechanistic account of life was examined more closely, and its inadequacy pointed out. This inadequacy is due to the fact that even by the widest stretch of imagination it is impossible to conceive of any mechanism capable of bringing about the phenomena, fundamentally characteristic for life, of organic maintenance and reproduction. The failure of the mechanistic account at this point is absolutely fatal to it, since the phenomena of maintenance and reproduction run through all organic activity.

Failure in the possibility of a physico-chemical theory of life implies failure in the physico-chemical account of Nature. The facts of observation are, moreover, altogether against the theory that the failure is due to interference within the living body of some non-physical agency, such as the supposed 'vital principle.' As soon as we

proceed to test this dualistic theory we find
that the supposed interference depends wholly
on what have been assumed to be physical and
chemical conditions, although the outcome of
these conditions is unintelligible from the
standpoint of physics and chemistry. To
understand the failure of the physico-chemical
theory of life we must examine the whole
physical and chemical interpretation of
Nature.

In the third lecture this examination was
undertaken in the light of philosophical
progress since Descartes first formulated
generally the mechanistic account of Nature,
including living organisms. The modern
world has become so accustomed to the
mechanistic interpretation of Nature that it
accepts this interpretation without question
as a representation of reality, and admits without question the claim of those who call
themselves 'realists' because they apply the
mechanistic interpretation logically in every
direction. But the claim was examined by
Berkeley, Hume, and other eighteenth-century
philosophers, and found to be baseless in
spite of all the practical utility of physics and

chemistry. The physical world is only a human interpretation of reality, and an interpretation which is, in ultimate analysis, not consistent with itself. Men still go comfortably along, with Dr. Johnson, in the old beliefs about 'physical reality'; but that does not concern the present discussion. Physical reality represents only a working hypothesis of limited practical application. We are therefore quite free to apply to the phenomena of life a more adequate working hypothesis.

We find this more adequate hypothesis already in practical use, but not clearly formulated, and, moreover, still confused by the intrusions of the mechanistic hypothesis. In the light of this more adequate hypothesis the parts and activities which we discern in a living organism and its biological environment can be understood only as manifestations of a whole which is the life of the organism, and which expresses itself in activity. The parts and their spatial arrangements, including environment, are the expression of active and unceasing maintenance, nutrition, or metabolism; and the activities are activities which express themselves in this

maintenance. Reproduction is only a phase
of that constant insensible reproduction which
is ordinarily called nutrition. If we examine
any living activity closely we find that we
cannot define its nature apart from its relation,
including spatial relation, to other associated
activities. The respiratory activity of the
tissues, for instance, is not something apart
from nutritional, secretory, or muscular
activity, but only one side of them; and the
structures or spatial arrangements involved
in respiratory activity are not something apart
from their activity, but an expression of
respiratory and other activity. If, for instance,
we forcibly prevent respiratory activity the
structures go to ruin. We can see at once
that this is so if we look closely at living
structure. The blood is part of the living
structure of the body, and at different parts
of the body the structure or composition of
the blood present locally is different. The
blood leaving the heart, or kidneys, or liver,
or muscles, or brain, is not the same in composition, but each variety is 'normal' for its
particular place, and is kept normal by
constant activity, just as is the rest of the

'structure' of these organs. The normality depends on regulation of flow in such a manner that normality is maintained. Thus the structure is the expression of activity, and this is so also for all the apparently solid structures of the living body, for the air in the lungs and the food-material in the alimentary canal, and for the external environment in so far as it participates in life. The structures or spatial arrangements and activities maintain themselves as a whole, and together constitute the whole which we call life. They cannot be understood as the separate parts and activities of a machine can be understood ; and the spatial arrangements of the parts, including environment, cannot be understood as those of a mechanical system. Life does not exist as something localised in space. Spatial relations are within it, as an expression of it. In presence of what we recognise as life we apply, and must apply, a special mode of interpretation differing completely from the physical interpretation which seems to suffice in the case of inorganic phenomena. The difference between organic and inorganic is for us a logical difference :

but we find as an experimental fact the logic of life before us when we study life, just as we have found up till now that the logic of mechanism is sufficient for the occasion when we study what we distinguish as inorganic Nature.   It is Nature we study when we study life, as Hippocrates first pointed out, just as it is Nature we are studying in physics and chemistry.   We cannot resolve life into mechanism; but behind what we at present interpret as physical and chemical mechanism life may be hidden for all we yet know.   This is an open scientific question; but the question whether life may not some day be resolved into physico-chemical mechanism is not an open question.   No possible meaning can be attached to such an expression as 'the mechanism of life.'

When we see that the distinctive conception of life is the foundation of biology we also see that biology is an independent and progressive experimental science advancing confidently like the physical sciences, and with a pedigree as good as theirs.   If, on the other hand, the view which has been popularised by so many physiologists in recent times were correct, that life is only physical and chemical mechanism, we could not say with any confidence that biology is progressing; for with every experimental advance we seem to be further from a physico-chemical conception of life.

In the present lecture conscious life, or personality, has been considered. An organism considered as a mere organism responds 'blindly' to the happenings in itself and its environment. For it they are mere happenings which it meets by blind efforts at maintenance and reproduction of its life. These happenings have no 'meaning' for it, so it does not learn from experience, and the happenings are just events arranged outside one another in time like the happenings to inorganic bodies. For personality, however, not only is the present, but also the past and the future, given in each happening. In perception we are aware of the past and future as well as the present; and voluntary action is action with both the past and the future in view, as well as the present. Even the simplest sensation carries within it a past and future; and more complex perceptions carry within them explicitly a past and future arranged in time and space and connected together by the logical principles through which our experience is constituted and bound together. Time, as well as space, are within personality. Both past and future shape themselves in the present. Personality does not exist in time, like mere life, and also does not exist in space any more than life does. The surrounding physical, chemical, and biological world is within personality : for the world is just the world of perception as Hume and Kant taught, or the world moulded and fashioned in thought and deed during progressive human development, as history and anthropology teach us.

When we examine personality more closely we find that just as the life of an organism is no mere individual life, but continuous without break from generation to generation, so personality is no mere individual personality among other personalities. We feel in, see in, exist in, that supreme Personality whom we call God, whose existence we recognise in the recognition of duty to our fellow-men and fellow-animals, present, past, and future : also to a truth which transcends practical truth for our individual selves, or for our generation.

We find also that personality manifests itself only in constant activity—in constant seeking for and acting on duty and truth. It is only so that we realise one-ness with God. The here and now, the imperfection, suffering, and sin of the world are not something outside of and indifferent to God. As the New Testament teaches, God is in this world, here within and around us, amid the ignorance, sin and suffering, and not apart.

We can now reach clear ideas as to the relations between time, space, mechanism, life, and personality. It is evident that the relation is a logical one, and depends upon different degrees of adequacy to reality in our conceptions of phenomena.

We may regard a man as simply a unit outside other units in time, as, for instance, in the counting of men in a division-lobby of the House of Commons. We may also regard a man as a mere extended patch of colour, or as occupying so much space. The artist who is arranging a picture, the soldier using khaki as camouflage, or the architect designing a suitable lift, regards him for the moment in the light of one or other of these very abstract conceptions.

We can also regard the man as about seventy kilogrammes of material with a certain external and internal configuration, and in more or less constant movement: such material consisting of a great variety of chemical molecules, acting upon one another and passing inwards and outwards in various ways. This is a much less abstract conception of the man, and for many immediate purposes is sufficient or more than sufficient. It is the account which the mechanistic theory of life aims at giving in a complete form.

We can next regard the man as alive—as an expression of blind maintenance of organic structure, activity and environment, continuous from generation to generation. This is the point of view of biology, freed from the abstractions of the mechanistic standpoint. This mode of regarding the man takes into account a great deal of what was entirely left out of account in the physical and chemical view of him. The biological account is evidently of enormous practical use, for instance in relation to medicine, agriculture, and social or economic efficiency.

Lastly we can regard the man as the expression of personality—of a reality which includes within itself the time and space around him and all that is within them. This reality manifests itself in his aspirations to see and do his duty—to discover the truth and make it prevail; and it is this reality which great teachers, poets, and artists are constantly revealing. As the expression of personality which is no mere individual personality, but enfolds experience, the man includes all that was left out in the previous pictures of him. He is thus no longer an abstraction from reality, like the man as a mere unit or form, or an arrangement of material parts, or a living organism, or a mere individual. He is now the real man, and his relations to the man as a material object, or as a living organism, is the relation of reality to an abstract and imperfect ideal presentation of it. Philosophy shows us that much of what, like the material man, is often taken for reality is only an insufficient ideal presentation of it. The materialists rather than the philosophers are those who mistake ideal constructions for truth, though Descartes made this mistake when he took clearness of theoretical representation as a criterion of truth.

Philosophy warns us against the mistake of supposing that a man is anything different from his organism perceived and understood more fully. It is absolutely vain to attempt to separate in any but a logical sense the personality of the man from his organic life or his physical and chemical mechanism. His character is, on a lower plane of perception, organic character : his passions on this plane are mere organic activities and instincts. To separate off the man's organic nature as if everything belonging to it were outside his spiritual existence is to make the same sort of mistake as is made by the vitalists in biology. The love between parent and child, or between man and woman, may be looked upon, by those who are not attempting to see further into reality, as mere organic instinct. In nothing else, however, does personality in its distinctive sense manifest itself more clearly, and through nothing else can character be judged more certainly.

Philosophy also warns us against mistaking the ideal physical interpretation of the univere for its reality. The astronomer seems at first sight to be presenting to us a gigantic and absolutely inhuman universe, in which man and human activity are but tiny specks in infinite space and time. The experimental physicist, on the other hand, seems to show us that we and our activities are only a fleeting appearance due to the equally inhuman whirling and clashing of myriads of electrons.

These are mere ideal interpretations, recognised as such by the great philosophers of the eighteenth century, but serving, nevertheless, to remind us, that man as a mere individual, or as a collection of them, is also a mere ideal interpretation. ' As for man his days are as grass : as a flower of the field so he flourisheth. For the wind passeth over it and it is gone; and the place thereof shall know it no more.'

The main object of these lectures was to define the concepts and aims of biology, and distinguish them clearly from the concepts and aims of other sciences : also to justify the claim of biology to rank as no mere branch of physics and chemistry but as an independent science of the utmost practical importance and of great philosophical significance.

Apart from philosophy, which looks at experience as a whole, we can never reach clearness about the knowledge dealt with in individual sciences; and this is why so much philosophical discussion has entered into these lectures. Just as an ordinary workman ought to understand his tools, their dangers, and what can or cannot be done with them, so ought a man of science to understand those tremendously powerful and dangerous logical tools with which he does the work allotted to him. Nothing is more wide of the mark than the contention that philosophy is not needed in scientific work. The inevitable result for men of science who ignore the history and results of philosophy is that they are far more apt to fall victims to all sorts of misunderstanding, and even to gross superstition. They are also apt not to see great scientific questions which are waiting for experimental investigation; and they are in danger of spending time fruitlessly on investigation which can lead nowhere. Those who imagine that science

deals only with ' objective facts,' independent
of the observer and the ideas he has inherited,
are the first to be misled.

In science we are always dealing with partial
and incomplete aspects of reality—with abstractions which are not only convenient but
ultimately unavoidable. Science is the application of abstract logical principles to a reality
which they can never express fully. This is
so not only in the mathematical, but also in
the physical, biological, and any one of the
humanistic sciences. It is the business of
philosophy to point out and define these
abstractions. Philosophy directs us, also, to
that spiritual reality which is the only reality ;
and from this point of view philosophy and
religion are one.